Vision and Change in the Geosciences

The Future of Undergraduate Geoscience Education

Sharon Mosher and Christopher Keane, Editors

Authoring Committee

Sharon Mosher, Jackson School of Geosciences, University of Texas at Austin, Chair

Wendy Harrison, Colorado School of Mines

Jacqueline Huntoon, Michigan Technological University

Christopher Keane, American Geosciences Institute

David McConnell, North Carolina State University

Kate Miller, University of Wyoming

Jeff Ryan, University of South Florida

Lori Summa, Rice University; retired ExxonMobil Upstream Research

Joshua Villalobos, El Paso Community College

Lisa White, University of California – Berkeley

Thank you to the Jackson School of Geosciences, University of Texas at Austin for its support for the events and activities needed to make this research possible.

ISBN-13: 978-0-922152-33-9

ISBN-10: 0-922152-33-0

Published and printed in the United States of America.

This material is based upon work supported by the National Science Foundation under Grant Numbers EAR-1347209, EAR-1725289, ICER-1748780, ICER-1740844 and ICER-1740386.

Any opinions, findings, and conclusions or recommendations expressed in this material are those of the author(s) and do not necessarily reflect the views of the National Science Foundation.

Design: Brenna Tobler

Copyedit: John P. Rasanen

Cover images, clockwise from top right: Samantha Berkseth, for AGI's 2015 Life as a Geoscientist contest; Victoria Heath, for AGI's 2017 Life as a Geoscientist contest; Issie Corvi, for AGI's 2017 Life as a Geoscientist contest; Riley Finnegan, for AGI's 2018 Life as a Geoscientist contest; ©Shutterstock/wavebreakmedia; Courtesy of the Jackson School of Geosciences, University of Texas at Austin; Courtesy of the Jackson School of Geosciences, University of Texas at Austin; Maxine Brown, image courtesy of the Electronic Visualization Laboratory at the University of Illinois at Chicago, for AGI's 2016 Life as a Geoscientist contest.

CONTENTS

Executive Summary

The ***Future of Undergraduate Geoscience Education* initiative**, sponsored by the National Science Foundation, addresses three critical questions facing undergraduate geoscience education:

- What concepts, skills, and competencies do undergraduates need to succeed in graduate school and/or the future workforce?
- What are the best teaching practices and most effective use of technology to enhance student learning?
- How do we recruit, retain, and ensure the success of a diverse and inclusive community of geoscience majors and support K–12 science teachers to contribute to a well-informed public and dynamic geoscience workforce?

Since 2014 over 1000 geoscientists in the academic and employer communities have collectively developed a robust vision for the *Future of Undergraduate Geoscience Education*. This report articulates that vision and identifies strategies for transformative change in undergraduate geoscience education. The key strategic findings are summarized below and highlight recommendations that capture the extensive work of the community participants and the initiative's organizing committee. These recommendations are comprehensive, and each department, program, or institution should consider how to appropriately implement them in the context of each institution's educational mission. Geoscience educators have an opportunity to capitalize on an evolving higher education landscape, the role that geoscience plays in addressing issues of importance to society, and changing demographics of the student population. We will lose that opportunity, however, if we insist on grounding our programs in the concepts, skills, and pedagogy of the 20th century.

KEY FINDINGS AND RECOMMENDATIONS

The academic and employer community participants have established a consensus on the broad geoscience concepts, skills, and competencies (i.e., expected student learning outcomes) that need to be developed to prepare graduates for meaningful employment across many occupations and/or for continued education in graduate school. Courses and activities should enable students to develop an understanding of geoscience concepts, including processes and impacts, that creates a strong framework for the future acquisition of knowledge. Activities that develop geoscience skills, building competencies at progressively higher levels, should be integrated into multiple classes so that students can practice, establish mastery, and recognize how these skills are employed. Quantitative reasoning and computational skills required in the workplace and/or

OBSERVATIONS FROM THE COMMUNITY:
Geoscience programs will grow and thrive when their graduates can demonstrate that their knowledge and skills are grounded in innovative thinking and have prepared them well for their role in a dynamic society.

Nick Gilbert for AGI's 2017 Life as a Geoscientist contest

graduate school should be incorporated throughout the curriculum. Professional skills (communication, teamwork, project management, etc.) should also be included in the undergraduate program along with those skills closely aligned with the discipline. The educational focus should be on students developing competencies that are embedded in a robust knowledge framework so that graduates can use these concepts and skills to solve geoscience problems throughout their careers. Authentic experiences, such as fieldwork, research projects, and exercises that involve real data acquisition and analysis foster geoscience habits of mind, systems thinking, and problem solving. Further, geoscientists need to combine critical thinking with a full understanding of accuracy, limitations, and uncertainty.

The community-vetted suite of concepts, skills, and competencies provides the basis for successful curriculum revision, in which student learning outcomes become the foundation of curricula planning. Geoscience faculty within a program and/or department should develop a consensus on course and curricular-level student learning outcomes, accounting for the recommended concepts, skills, and competencies and institutional priorities and capacities. Strategies such as backwards design or similar approaches allow evaluation of how and whether targeted learning outcomes are being met by the current curriculum and can guide redesign of individual courses and/or the curriculum as a whole.

Student learning outcomes for discipline-specific and professional competencies can form the basis for geoscience program evaluation protocols. Each program should develop their own understanding of what defines their program's success, including effective and tractable learning assessments based on key student learning outcomes that incorporate disciplinary, institutional, and program-level goals. Many external assessment instruments and methods are available, such as those by the American Association of Colleges and Universities and *National Association of Colleges and Employers*.

Geoscience educators should further embrace and become adept at active teaching strategies that pedagogical research has shown motivates students and improves learning. Experiential learning courses and inquiry-based activities in laboratory courses emphasize the process of scientific discovery and promote a focus on students' roles in investigating scientific questions and building conceptual understanding. Using current and emerging technology and computational models and simulations with large datasets will increase student understanding of complex geologic structures, features, and spatial relationships and provide insight into processes and global-scale events.

The growing demand for geoscientists and the importance of increased workforce diversity requires programs and departments to recruit, retain, and ensure the success of a diverse and inclusive community of undergraduate geoscience majors. A geoscience community that pulls from the greatest breadth of society will gain from diverse life experiences and perspectives that capture unique insights and solutions to geoscience-related problems facing society. By engaging the entire student population, the geosciences can tap a greater range of talent and compete for the best minds.

A crucial first step to increasing geoscience enrollments is significantly improving the public's perception of the geosciences, by promoting it as highly relevant to societal and environmental issues and an economically viable, innovative career. Programs and departments should develop positive recruitment programs for new students, lower division non-majors, transfer students, and students underrepresented in the geosciences, leveraging existing institutional recruitment efforts. Another approach to increasing participation in Science, Technology, Engineering and Mathematics (STEM) majors in college is developing or collaborating with programs for minority students at pre-high school and high school levels to build the pipeline.

Programs should be intentional about the retention and success of recruited students, as part of a broad diversity, equity and inclusion (DEI) plan. Best practices include mentoring, building community, and other supportive actions, with particular focus on students underrepresented in the geosciences. Establishing programs that support students before, during, and after transferring from community colleges to four-year programs promotes student success and can help in increasing enrollments and diversity in four-year institutions. Additionally, geoscience programs can leverage institutional efforts to build partnerships among geoscience programs at two-year colleges, four-year colleges/universities and minority-serving institutions to enhance diversity, equity and inclusion in the geosciences.

Introductory and non-major courses should leverage the Next Generation Science Standards (NGSS) to engage all students and preservice teachers to align the courses with students existing expectations of how science is taught. Programs and departments should revise these courses to focus on processes and systems, integrate other sciences and math, and use active-learning pedagogies and resources. This approach may entice more students who take these courses to major in the geosciences and will prepare proficient geoscience-literate K–12 teachers. Explicitly identifying and using cross-cutting science and math examples in introductory and non-major courses demonstrates the relevance of the geosciences to students, including pre-service teachers for whom an introductory or non-major geoscience course may be their only science course. Societally relevant examples have the potential to be adapted for subsequent use in teachers' classrooms. Faculty teaching these courses should have the opportunity to participate in professional learning experiences that introduce the NGSS and supporting practices. If institutionally appropriate, geoscience departments should consider developing an option for K–12 preservice teachers, which may increase future college-level geoscience enrollments.

Faculty, advisors, programs, and/or departments should guide students to be proactive in their education and support them in identifying co-curricular opportunities for developing skills needed for future careers. Advisors and mentors can help students build customized roadmaps for attaining their educational and career goals. Additionally, students should have access to information on career options and training on how to find and obtain employment through institutional career centers, employers and/or alumni, professional societies, and other professional development resources.

Both geoscience faculty and students should recognize that formal undergraduate education is a robust foundation for lifelong learning in support of a successful career. Geoscience programs can help students become aware of external certifications required for some geoscience employment and the availability of continuing education programs. Moreover, students need to be prepared for changing workforce needs, including

Courtesy of the Jackson School of Geosciences, University of Texas at Austin

new careers and jobs that require the use of new technologies, strong quantitative and computational skills, data analytics and machine learning, interdisciplinary teamwork and problem solving.

An ongoing initiative investigating the skills and competencies needed by graduate students for successful careers in Earth, ocean, and atmospheric sciences underscores the need for undergraduate programs to build the educational foundations essential for students who will pursue graduate degrees. Most identified skills and competencies are similar, requiring a higher level of accomplishment, along with some additional skills related to research, problem solving, and future technological trends. Student success can be strengthened by supporting the career goals of students and working with them to build the competencies they need for the future.

Case studies captured by this initiative demonstrate that efforts to revise undergraduate programs are multi-year processes requiring patience, persistence, and steady leadership to maintain engagement and sustain momentum. Heads and chairs need to encourage, facilitate, and support those faculty tasked with making changes to undergraduate programs; they need to allocate necessary resources, assure alignment of curricular efforts with institution-level priorities, and keep the upper administration informed of progress in these activities and of the national effort that necessitates these changes. Review, revision and changes to undergraduate programs and teaching are best accomplished through bottom-up efforts and identifying key individuals or teams of faculty to drive the effort while maintaining full transparency with the rest of the department. Successful implementation of active and experiential learning and other research-based strategies that improve student learning and motivation are greatly facilitated by professional development opportunities and other incentives for faculty, such as release time, redistribution of workload, or reduction of non-instructional assignments, as allowed by each institution.

A wide range of stakeholders, including academics, employers, and organizations, have vested interests in the success of undergraduate geoscience education and bear a responsibility for accomplishing this vision for the future. Heads and chairs should encourage and support faculty in completing necessary curricular review and revision efforts and in adopting new instructional approaches. They should also advocate to their Dean for support of these efforts related to meeting community educational standards and articulate their relationships to institutional measures of student success. Faculty should participate in and leverage relevant professional development experiences, the more specific to the geosciences and to their courses the better. Funding agencies, professional societies, and other stakeholder groups should support and/or offer a broad array of relevant professional development experiences for faculty to learn how to adopt innovative teaching practices and understand curricular enhancement and revision efforts. Academic departments and geoscience employers should establish and maintain interactive professional relationships with one another, focusing ultimately on improving the abilities and accomplishment of bachelor's level geoscience graduates.

Sustained change in geoscience undergraduate education will require the combined and coordinated efforts of departments and programs, administrators, individual faculty, geoscience employers, and professional societies. To prepare geoscience undergraduates for success requires cultural change, from the administration down to the students. This report is a roadmap and resource for deans, heads and chairs, undergraduate program directors, faculty, other educators, current and future employers, and professional societies in shaping the future of undergraduate geoscience education.

Courtesy of the Jackson School of Geosciences, University of Texas at Austin

Annotated Table of Contents

Courtesy of the Jackson School of Geosciences, University of Texas at Austin

Courtesy of the Jackson School of Geosciences, University of Texas at Austin

8. **Recruiting, Retaining, and Promoting Success of Undergraduate Students with Emphasis on Students Underrepresented in the Geosciences**
   Summarizes best practices compiled from Summit participants and other community advice

9. **Leveraging the *Next Generation Science Standards* in Introductory and Non-Major Courses to Recruit Majors and Prepare K–12 Science or Geoscience Teachers**
   Outlines the impact of the *Next Generation of Science Standards* on undergraduate education and how to use it for recruiting majors and preparing the next generation of K–12 teachers

10. **Building a Learning Continuum for Life — Undergraduate Geoscience Education as Part of the Continuum of Career and Lifelong Learning**
    Discusses the importance of and how best to accomplish lifelong learning

11. **What Graduate Geoscience Education Should Accomplish**
    Summarizes the outcomes of the 2018 Geoscience Employers Workshop for graduate education in Earth, atmospheric and ocean sciences

12. **Fostering Change in the Academic Community: Case Studies**
    Provides results of heads and chairs efforts to implement change using the information outlined in this report

13. **Engaging Stakeholders**
    Discusses what stakeholders need to do to accomplish change to undergraduate geoscience education to prepare students for the future

1. Introduction: A Call for Action

During the last six years, over 1,000 members of the geoscience community have helped shaped a vision for the future of undergraduate geoscience education to better prepare future geoscientists. This vision provides a roadmap as educators to make critical, positive changes to our programs over the next decade. This call for action is motivated by many factors ranging from the success of our students to the health of the geoscience profession.

Nicholas Perez for AGI's 2017 Life as a Geoscientist contest

As educators, we need to help students build a foundational understanding of how the Earth system works, so they can apply this knowledge to complex and profound issues where geoscience and society intersect. Evidence of geoscience processes impacting society is increasingly common and highlight geoscience-themed "grand challenges" that will confront our global population in the coming decades, including more frequent and intense natural hazards, threats to coastal infrastructure from rising sea levels, the effect of our warming climate on food and water supplies, and meeting global energy and resource needs in a sustainable and environmentally responsible way. The types of thinking inherent to the geosciences — working with complex systems, temporal and spatial reasoning, and the collection, interpretation, and analysis of complex natural data — are among the most crucial approaches to addressing these future challenges.

Geoscience research has evolved to meet these challenges, with interdisciplinary, multi-disciplinary and transdisciplinary teams addressing complex Earth system problems of societal importance with new tools, technologies, and approaches. Both undergraduate and graduate education need to undergo similar changes to prepare students for their future. Successful geoscientists need a strong foundation of disciplinary knowledge and scientific skills while being able to work across domains.

Beyond academia, employers are also responding to a new reality that includes use of multidisciplinary approaches by diverse teams drawn from a broad range of backgrounds, experiences, and capabilities. These teams use innovative tools, models, and skills to integrate new ways of thinking and a range of datasets. Recent geoscience graduates are entering fields that have experienced rapid changes in how work is done, due both to significant changes in the type of work and skills needed and from the incorporation of new technologies. Employment opportunities for geoscientists are expanding. Graduates with training in economics, risk management, ethics, or policy, and the ability to work across cultures and communicate with diverse audiences will have a broader array of opportunities for career advancement. Some future graduates may find

Courtesy of the Jackson School of Geosciences, University of Texas at Austin

themselves in occupations that did not exist a decade ago, while others will find themselves in occupations that are yet to be invented.

Our understanding of the complex interactions between different parts of the Earth system, including the Earth's interior and surface, hydrosphere, atmosphere, cryosphere, and biosphere, and in the chemical, physical, biological and geological processes that help shape the Earth system continues to grow. Yet much of undergraduate education still compartmentalizes these processes and components. Geoscience programs need to adapt curricula to emphasize these critical interactions. While the geoscience community has begun to create resources that support interdisciplinary learning about the Earth, and the important role that the geosciences play in addressing grand challenges, we have much more to do to incorporate these ideas deeply into geoscience programs.

Student success requires more than just sophisticated content knowledge. The geoscientist of the future needs a range of skills and competencies. Students who can conduct quantitative analysis, apply critical thinking and problem-solving skills, manage and analyze large data sets, communicate effectively in a variety of formats, and work well in teams will be likely to succeed in the future work environment, even if they choose non-geoscience employment. The knowledge and skills needed for success will change throughout a career, so creating programs that encourage students to develop the flexibility to adapt to change will be important. Geoscience departments must assess their programs and incorporate opportunities to facilitate the development of appropriate skills and competencies.

How undergraduate education is delivered by faculty to students has also been transforming. We now understand that HOW we teach has a direct impact on how well our students learn. Effective, research-supported teaching strategies have been increasingly incorporated in STEM classes during the last several decades but remain far from ubiquitous. Widespread adoption of these approaches to teaching by geoscience college instructors will lead to students who are more successful and better-prepared to enter the workforce.

Education research tells us that students learn better when they can actively monitor their understanding with purposeful activities *during class*. Furthermore, knowledge is socially constructed, and people learn best in supportive social settings such as small collaborative groups in active learning environments. Active and experiential learning increases student learning and reduces attrition in science, math, and engineering courses and can reduce performance gaps among different student populations, especially those that are traditionally underrepresented in Science, Technology, Engineering and Mathematics (STEM) disciplines. We must support improved student performance and develop more compelling learning opportunities by reshaping the way we approach university teaching and learning and adopting documented best practices.

As our understanding of effective practices in undergraduate education has grown, technology has become more readily used to support learning, including virtual and interactive experiences, videos and animations, blended learning modalities, and crowdsourcing of open education resources. These advances enable multiple types of opportunities for learning and the sharing of community created educational resources. Furthermore, significant advances have occurred in tools for visualization and geospatial analysis, for using massive amounts of quantitative information, and for computational modeling, animation, and simulation for both predictive capabilities and generating new insights into Earth processes and global-scale events. Undergraduate students need to be prepared to use these rapidly evolving technologies and the large data sets they produce in the future.

As a substantial segment of the current workforce retires and geoscience opportunities increase, we will face shortages of geoscientists in the future workforce if enrollments do not increase. Despite continued efforts by educators and industry and with considerable support from federal funding agencies, the geoscience community still has difficulty recruiting students from underrepresented groups to our programs and sustaining success through completion of their degree and into the profession. Changing how geoscience departments and programs in 4- and 2-year institutions recruit, mentor, support, and educate students is necessary to increase enrollments and ensure that we are creating an inclusive and diverse workforce.

The Next Generation Science Standards for K–12 education have put Earth and Space Sciences on an equal footing with Physical and Biological Sciences, which is a tremendous opportunity and challenge for our field. Using the NGSS approaches to teaching science will help us improve introductory and non-major courses and align them with learning expectations of new students. Additionally, this is a pivotal time to engage pre-service K–12 teachers in our undergraduate programs and courses to build their content knowledge, scientific skills, and confidence in facilitating students' investigations into Earth science questions in their future classrooms. Geoscience content in middle or high school courses can create an interest in geoscience careers while contributing to development of a scientifically literate society.

Over the last decade, the expectation that STEM graduates will be prepared for the workforce has increased. Calls for change to undergraduate education have come from the National Academies and National Research Council, the Association of American Universities, Association of American Colleges and Universities, the National Science Foundation, the White House Office of Science and Technology Policy (OSTP) and the President's Council of Advisors in Science (PCAST). Legislatures, governing boards, regents, presidents, provosts, as well as parents, alumni, the public and students have put concurrent widespread pressure for change on universities. This concern is fueled in part by an increased divergence between the needs in the workforce and what undergraduates learn. A focus on concepts and skills, and the ability to use them (i.e., competencies) provides an excellent educational foundation while preparing students for future success regardless of their career path.

Departments are more likely to attract students and thrive if they invest their time and resources in helping students develop a rich set of critical skills and content knowledge clearly linked to potential careers and the aptitude to manage their developmental pathway. Instructors and students are investing in the multi-year development of life-long skills, and the return on that investment comes when students secure employment and are prepared for a successful career in a wide range of occupations and work settings. As educators, we derive personal and professional satisfaction when our students succeed, and we have an obligation to prepare our students for the future.

Christopher M. Keane

The community vision outlined in this document provides a roadmap for positive changes to undergraduate geoscience education, recognizing the differing priorities and capacities among our diverse educational institutions, and promoting strong collaborations between educational institution types (e.g., community colleges and four-year institutions). Lasting changes for the benefit of our profession and the students we educate will take the combined efforts of geoscience departments and programs, led by administrators, individual faculty, geoscience professional societies, and employers.

Our learning environments and curricula must evolve to address future geoscience challenges and prepare students to build long and successful careers. Given regional and global challenges, the accelerating pace of change in all aspects of society, a renewed commitment to diversity, equity and inclusion, and the evolution of geoscience careers, geoscience educators urgently need to reconsider our role in educating the next generation of geoscientists and in producing geoscience-informed graduates generally.

Courtesy of the Jackson School of Geosciences, University of Texas at Austin

KEY OUTCOMES

- The academic and employer communities have developed a consensus on the broad geoscience concepts, skills, and competencies that need to be developed throughout the curriculum across multiple courses and educational experiences.
- The community-vetted concepts, competencies, and skills provides the basis for successful curriculum revision in which students learning outcomes become the foundation of curricula planning.
- Geoscience educators should further embrace active teaching strategies that research has shown to improve student learning.
- The community-vetted suite of student learning outcomes for discipline-specific and professional competencies can be used as the basis for geoscience program assessment.
- Departments should help students take a proactive role in their education and co-curricular solutions to develop skills needed for future careers.
- Growing demand for geoscientists requires departments and programs to recruit, retain, and promote the success of undergraduate geoscience majors across a broad spectrum of society.
- Introductory and non-major courses should leverage the Next Generation Science Standards to engage all geoscience students and preservice teachers.
- Programs and students must recognize that formal undergraduate education is a robust foundation for lifelong learning in support of a successful career.
- Skills and competencies needed by graduate students for successful careers should be integrated into Earth, Ocean, and Atmospheric Sciences graduate programs.
- Undergraduate program revision efforts are multi-year processes that require patience, persistence, and leadership to maintain engagement and sustain momentum.
- Many individuals and organizations have a stake in the success of undergraduate geoscience education and a responsibility to help accomplish the vision for the future.

2. Process Summary: Summits, Workshops, Survey

The Future of Undergraduate Geoscience Education initiative, sponsored by the National Science Foundation (NSF), addressed three critical questions facing undergraduate geoscience education:

- What concepts, skills, and competencies do undergraduates need to succeed in graduate school and/or the future workforce?
- What are the best teaching practices and most effective uses of technology to enhance student learning?
- How do we recruit, retain, and ensure success of a diverse and inclusive community of geoscience majors and support K–12 science teachers to contribute to a well-informed public and dynamic geoscience workforce?

In 2014, the *Summit on the Future of Undergraduate Geoscience Education* made major progress developing a high-level community vision for the geosciences. The Summit brought together a broad spectrum of the undergraduate geoscience education community, about 200 faculty from Carnegie Classification R1, R2, and R3 research universities with undergraduate programs, doctoral/professional universities, terminal Master's programs, four-year private and state colleges (4YC), and 2-year community colleges (2YC) from across the country, as well as about 20 representatives from industry, government and professional geoscience societies.

Energized by keynote presentations and informed by panel discussions, the participants spent nearly three days discussing the three questions above in small working groups. The working groups then presented their summarized results to all Summit participants, prompting discussions among all Summit participants. A key point of consensus from the working group reports was that developing critical competencies, skills, and conceptual understandings was more important than students taking a specific menu of courses. Attendees agreed to a list of important conceptual, scientific, and geoscience specific skills and competencies (Figs. 3-1, 3-2, 3-3), and a summary of the Summit outcomes was published on the web (Mosher et al., 2014).

A comprehensive electronic survey (Appendix A) was distributed nationally to geoscience faculty and employer contacts following the Summit. Approximately 470 individuals responded: 77% academics and 23% employers (17% industry, 3% government, 2% other, 1% professional societies). Approximately 85% were not Summit participants. Although the survey remained open through 2019, most responses were submitted in 2014 and 2015. The survey respondents agreed with the 2014 Summit participants that improving competencies, skills, and conceptual understanding was the most important issue in undergraduate

Irish Centre for Research in Applied Geosciences (iCRAG)

education, with 95% of the survey participants rating this goal as very important or important. Eighty percent of the survey respondents agreed with the major conclusion of the 2014 Summit that developing competencies, skills, and conceptual understanding was more important than taking a specific menu of geoscience courses. Respondents also concurred with the list of important concepts and skills generated at the Summit.

Additionally, the survey included questions on the status of curriculum reform efforts in departments, on the use of various research-validated teaching methods, on the constraints on geoscience degree programs, and, for both the departments and employers, on the type of programs offered or supported for broadening participation of underrepresented groups and for K–12 teacher preparation in geosciences.

The *2015 Geoscience Employers Workshop* brought together 46 participants evenly distributed among representatives of the energy industry, hydrogeology, engineering and environmental consulting firms and companies, and various government agencies. One representative from a mining company was present, along with a few professional society representatives. This workshop further investigated the important concepts, competencies, and skills students need for success in the future workforce. The workshop followed the same format as the *2014 Summit*, with working groups comprised of employers from different sectors focused on topics related to universally-needed skills and concepts. The geoscience employers strongly agreed with the *2014 Summit* and *2014–2015 survey* outcomes on the skills and concepts undergraduates need, regardless of employment sector (see Summa et al., 2017). In addition, they provided increased granularity on key skills and concepts and identified that in their view experiential learning strategies as the best way to instill these skills in students (Appendix B). They also provided insights into how industry could help departments implement this community vision.

In 2016, a second Summit event was held specifically for department heads, chairs, and other administrators. This Summit focused on further development of the emerging community vision for undergraduate geoscience education and to develop implementation strategies for this vision at the departmental level. Participants included over 100 geoscience academic leaders from Carnegie Classification R1, R2, and R3 research universities with undergraduate programs, doctoral/professional universities, terminal Master's programs, four-year private and state colleges (4YC), and 2-year community colleges (2YC) from across the country. The *Heads and Chairs Summit* followed the same format as the previous events, with the additional assignment that each institution submit an action plan for their department.

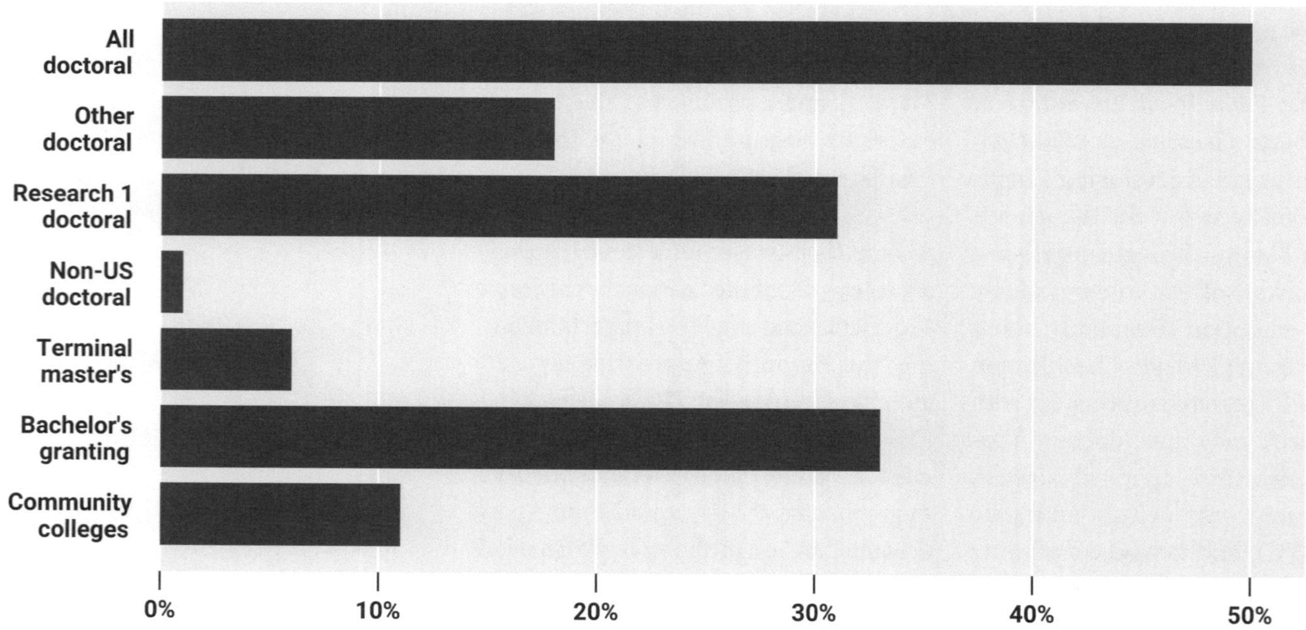

Figure 2-1: Contributed Input Levels by Institution Type
Inputs from academia

These development plans varied widely depending on size and type of institution, stage of undergraduate curriculum evaluation, and areas of focus (i.e., curriculum, pedagogy, broadening participation, etc.). Several follow-up workshops targeting Heads and Chairs were held between 2017 and 2019 to increase the number of departments engaged in the process and to gather further input and departmental action plans and progress reports. By 2017 a total of 91 individual department action plans were submitted. Progress reports on these action plans were submitted by 56 departments, gathered between 16 to 36 months after the *Heads and Chairs Summit or 2017 Earth Educators Rendezvous* workshop; 12 submitted a second follow up progress report providing insight into the time horizon for needed change. The results of these reports are discussed as case studies in Section 13 and presented and analyzed in Appendix C. These reports provide insight on best practices for curriculum reform, changing pedagogy, and other undergraduate program changes based on the experiences of the responding heads and chairs. In addition to the case studies in Section 13 and Appendix C, we have interspersed quotes from these progress reports where relevant throughout this document.

In 2018, as part of a new NSF-sponsored initiative on *Improving Geoscience Graduate Student Preparedness for the Future Workforce*, we held a second *Geoscience Employers Workshop*, representing a broader range of employment sectors and disciplinary specialties. The purpose was to explore the universal skills and competencies that should be part of graduate education for doctoral and master's students in Earth, ocean, and atmospheric sciences. Although this workshop concentrated on the graduate experience, geoscience employers also discussed skills they expected undergraduates to develop before coming to graduate school. Participants in this workshop highlighted the same menu of skills and competencies identified in the undergraduate effort, but the increase in importance of some skills was notable.

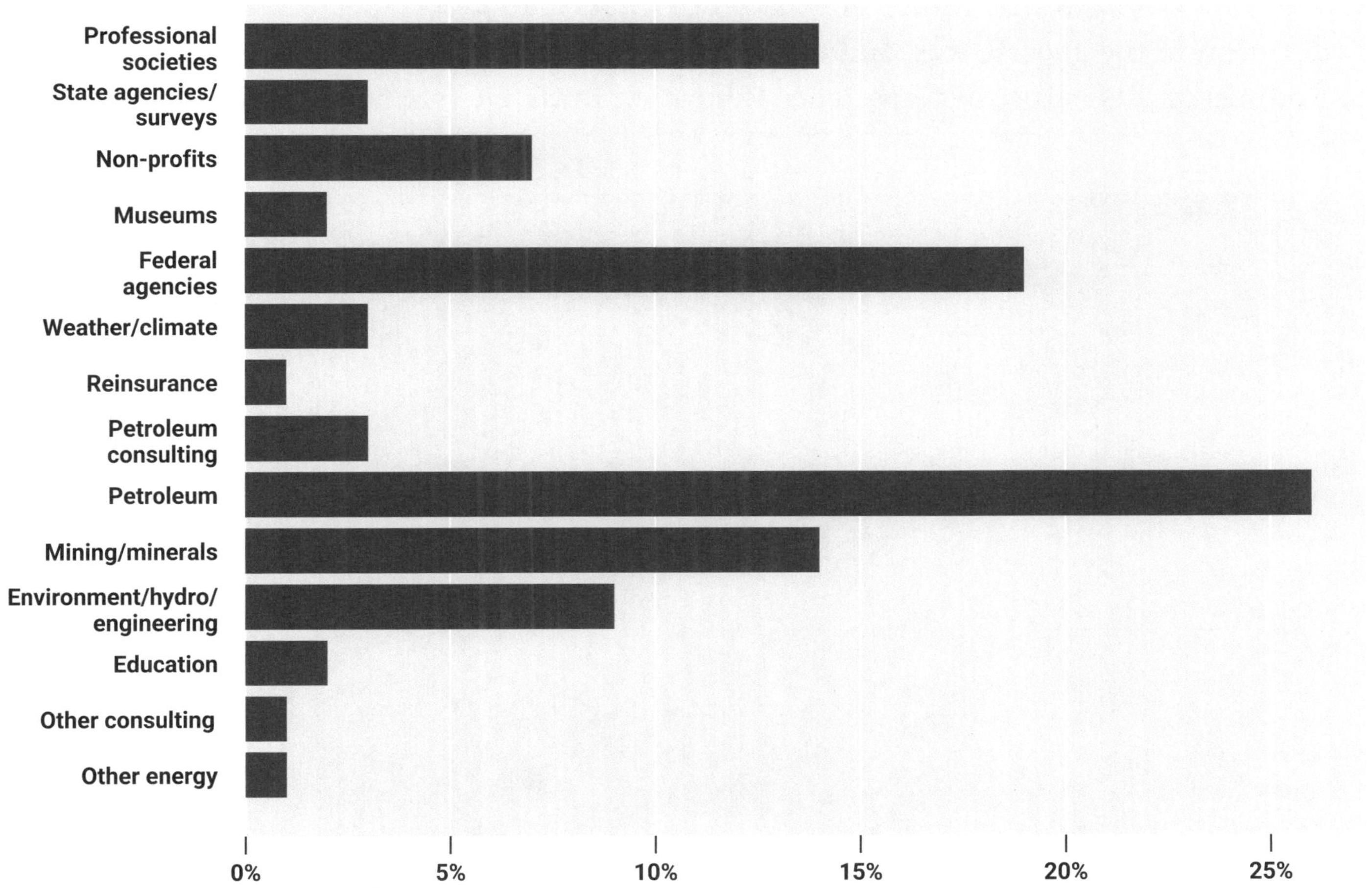

Figure 2-2: Contributed Input Levels by Employer Type

Sources of employer input

Overall, the academic and employer community involved in this geoscience education initiative reflect the distribution of faculty by institution type and of geoscience employers (Fig. 2-1 and Appendix D). Input from academic institutions was: 11% two year colleges (2YC), 33% Bachelors, 6% from M.S., and 50% Ph.D. (31% R1s, 18% other Ph.D., 1% international) granting departments. Input from employers was: 29% petroleum related, 19% federal agencies, 14% mining/minerals, 9% hydrogeology/environmental/engineering consulting, 4% museums/other education, 3% state agencies/surveys, 3% from weather/climate related, 1% energy, 1% reinsurance, 1% other consulting, 7% NGOs and 14% professional societies. For the undergraduate effort, most the participants were from the solid earth geosciences with some representation from the ocean and atmospheric sciences. For the graduate effort, there was an even distribution of participants across Earth, atmospheric, and ocean sciences.

An initial draft of this document was written in mid-2019 with each member of the committee writing sections most closely aligned to their expertise based on findings from the Summits, followed by a review of all sections by all committee members. The changes and additions from individual committee members were compiled by Mosher and Keane and the document reworked to have a consistent style. In mid-2020, the committee met to discuss the new draft and construct specific recommendations. After modifications, the committee reviewed the entire document again and final edits were incorporated. The revised document was reviewed by four external reviewers and based on the reviews further revised by Mosher and Keane.

This document outlines a robust community-wide vision supported by both representatives of the geoscience academic programs and employers. Well over 1,000 geoscientists have provided input to this vision for the future of geoscience undergraduate education. Additional studies of education in the geosciences, and of STEM education in general, support and substantiate the findings of this Future of Geoscience Education initiative (see references).

NOAA/Nicholas Morgan

3. What Undergraduate Geoscience Education Should Accomplish

The academic and employer communities have developed a consensus on the broad geoscience concepts, skills, and competencies that need to be developed throughout the curriculum across multiple courses and educational experiences.

Victoria Heath for AGI's 2017 Life as a Geoscientist contest

The geoscience academic community overwhelmingly agreed at the *2014 and 2016 Summits* that improving competencies, skills and conceptual understanding was the most important issue for undergraduate education. Similarly, 85–87% of the *2014–2015 survey* participants agreed that was very important and an additional 6–10% rated it as important. The major conclusion of the *2014 Summit* that developing competencies, skills, and conceptual understanding in students was more important than taking specific undergraduate courses was also validated by the *2014–2015 survey* with 81% of academics and 77% of employers agreeing. Consequently, the focus of this section is on concepts (A), skills (B), and competencies (C).

A. CORE CONCEPTS FOR SUCCESS

At the *2014 Summit* and in the *2014–2015 survey*, the academic community identified broad concepts that undergraduate geoscience majors, mainly geologists and geophysicists, should know to be successful in graduate school and the workforce. These broad concepts, ranging from core topics taught in most programs to those covered to a variable extent, are:

- Earth as a complex system
- Deep time
- Surface processes
- Earth materials
- Earth structure
- Hydrogeology and water science
- Natural resources
- Climate change

Both academics and employers concurred on the importance of these concepts (Fig. 3-1) (Appendix A). The *2015 Geoscience Employers Workshop* participants provided even further granularity on what they thought students should know and be able to do. Many of these concepts emphasize ideas taught in traditional courses offered by most departments (such as deep time, structure, Earth materials, surface processes), while others are covered only by some programs and only to variable extents (climate, hazards, resources, hydrogeology). The employers focus was on developing an *understanding* of broad concepts — i.e., the processes and their impacts — to build a working framework for knowledge gained during their education and future career. Although the employers provided a robust discussion of what comprised each broad concept as outlined below, they deemphasized the importance of memorization of terms, definitions, classifications, and other material that is easily referenced online. The employers at the *2015 Geoscience Employers Workshop* and participants at the *2016 Summit* did not address geologic hazards as a separate concept, but instead integrated hazards into the other appropriate concepts, which is reflected in the following subsections.

1. Earth as a Complex System

Objective: Students will be able to interpret Earth's systems, their interrelationships, and effects on each other.

Earth is a nonlinear, complex system composed of interlinked dynamic systems of different complexities of scale and interactions. These systems involve the movement of energy, solids, and fluids, commonly in cycles with different residence times. The systems and their parts interact and form feedback loops. Students need to understand the concept of systems and how they work. They must think of Earth as a system and understand its constituent interactions and forcing mechanisms. They need to consider the implications of forcings within the system and evaluate predictions. Students should particularly understand the human drivers of change and its impacts, environmental transitions, and scales over which changes occur. Students should be able to use present processes to infer past processes and use the geologic record of past processes to forecast the future, assessing the risks and advantages to changes in the Earth system.

2. Deep Time

Objective: Students will be able to comprehend the scale and magnitude of deep time, the impact of different time scales on geologic processes, and interpret its relationship to major geological and biological events in the geologic record.

Understanding geologic time is fundamental to all geosciences. Conventional means of determining relative time (superposition, cross cutting relationships, fossil succession, etc.) and absolute time (radiometric dating) should be understood, along with the precision and limitations of each of these methods. Students should be able to extrapolate from what is determined in the lab to the field and integrate this information into real geologic settings. Students need to understand the impact of time on different geologic processes. Processes can occur in real, instantaneous time, over millennia, or over millions to billions of years. The relevance of these processes depends on these different time scales. When examining Earth events, it is important that students consider rates, rates of change, duration, frequency, magnitude, residence time, scale, sequencing of events, and other impacts of time on geological processes. Examples where time is important include weathering and erosion, plate tectonics and the resultant natural hazards, the development of mineral and energy resources, and environmental changes and related hazards. Students must be able to employ temporal reasoning to address geologic problems and should recognize

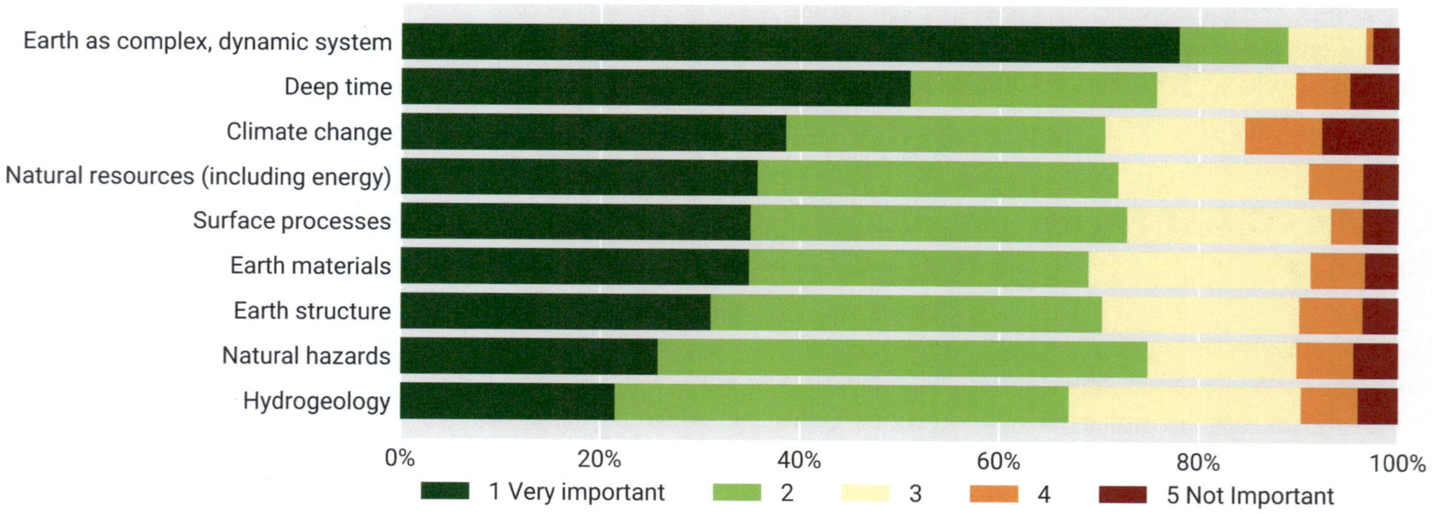

Figure 3-1: Geoscience Concepts Importance

Survey Question: The Summit report identifies the following concepts as critical to undergraduate education. For each, please indicate the importance from your perspective of these concepts for undergraduate-level curricula.

that different processes were more or less critical during specific periods in geologic history.

3. Surface Processes

Objective: Students will be able to recognize key surface processes and their connection to geological features and possible natural and man-made hazards.

The Earth's surface, in both the terrestrial and marine realms, is modified by physical, chemical, and biological interactions with variable rates of physical and chemical changes. Students should understand sediment deposition and erosion processes, including transport relationships and the magnitude and frequency of surficial deposits. Subsurface analogs that cannot be measured directly should be introduced along with field exposures. Particularly important are stream and river flow and the resultant morphologies, associated deposition and erosion, and the effects of floods. Streams and rivers are the primary mechanism that shapes the continents in our current climate, a major hazard, and an important resource for humans for water, hydropower, and recreation. Familiarity with landscape alteration (or geomorphology) through mechanical and chemical processes is important, including the development of karst, glacial deposition and erosion, landslides, and wind-related surface processes. Students need to make the connections between surface processes and the Earth's habitability and ability to sustain life. Surface processes are involved in natural hazards that affect society, and students should understand and evaluate the triggers, impacts, and risks associated with these processes.

4. Earth Materials

Objective: Students will be able to analyze and interpret the chemical and mechanical processes that are involved within each stage of the Rock Cycle.

The Earth is composed of rocks comprised of minerals with different physical and chemical properties. Students should know what rocks and minerals are, understand the rock cycle, and be familiar with the mechanical and chemical characteristics of rocks and minerals that determine how terrestrial solids behave under different conditions and change over time. Students should also know how physical and chemical properties of rocks and minerals are measured, the relevant scales of measurement, and the natural scales of heterogeneity. Knowing the processes that form rocks and minerals, the conditions of rock formation, fluid flow, dynamics and chemistry, and the role of microorganisms on rock formation are equally important. Students should recognize the role of rock formation in the localization and development of mineral deposits and energy resources, in terms of both organic and inorganic processes.

5. Earth Structure

Objective: Students will be able to analyze and interpret plate tectonic and deformation processes, the relationship to Earth's structure, and the resultant geological structures and natural hazards.

Plate tectonics is the central dynamic concept of the Earth system. Students need to know the different types of plate boundaries, their associated geologic and

Michael Collier, from the AGI ESW Image Bank

geophysical features, the driving forces of plate tectonics, and how these forces interact with other parts of the Earth system. Students must be able to link plate tectonic processes to deformation, basin formation, earthquakes, tsunamis, volcanism, and other hazards. Also important is recognizing the episodic nature and uncertainty of plate tectonic processes as they impact hazard prediction, mitigation, and planning. Central to understanding the dynamic Earth is knowing the internal layered structure of the Earth, the mechanical properties and composition of the different layers, and the various means by which we measure and define the Earth's internal structure (seismic waves, analysis of earthquakes, analogies to meteorites, etc.). Additionally, geoscience students should understand stress and strain, rock mechanics, deformation processes, and the conditions under which different kinds of deformation occur. Students should be able to recognize the resultant structural features, including folds, faults, shear zones, fractures and joints, and know the conditions under which they form. Students should also understand the importance of structural controls on energy and mineral resource accumulation.

6. Hydrogeology and Water Science

Objective: Students will be able to analyze and interpret the chemical and mechanical processes that are involved in the Water Cycle and important aspects related to water being a natural resource.

The water cycle is a critical component of the Earth system that impacts all life and the Earth's habitability. Students should examine the interactions among the hydrosphere, atmosphere, and land surfaces and among oceans, ice, fresh surface water and groundwater. Understanding ocean chemistry, dynamics and its layered structure is important. Students should know about groundwater and aquifers, including the difference between confined and unconfined aquifers, saturated versus unsaturated conditions, water phase behaviors and scales of aquifer heterogeneity in space and time. Additionally, students must recognize the importance of aqueous geochemistry and biogeochemical phenomena in the hydrologic cycle, including microbial interactions and nutrient cycling, contaminant transfer, groundwater quality, and impact on oceans. Discussions of economics, regulatory standards, and public policies regarding groundwater should be included.

7. Natural Resources

Objective: Students will be able to relate the distribution of natural resources to geological processes, explain how natural resources are formed, used, and extracted, and understand their relative availability.

Natural resources play a critical role for society, and it is important for students to understand the economic geology of natural resources, including water, energy, minerals, and other geologic materials. This understanding includes the economic importance of natural resources as commodities with finite availability, the economics and viability of different resources, and the distinction between resources and reserves. The geographic distribution of resources, and the geologic processes involved in their formation, extraction, and use, should be part of the geoscience curriculum. Students should have the opportunity to evaluate the pros and cons of renewable and non-renewable (finite) resources, investigate resource dependency and limits, know where materials, energy and medicines come from and how they are made, processed, or refined. Additionally, students should understand ecosystems services and their direct and indirect contributions to human well-being. Students should be familiar with the relative supply and demand for natural resources, including ore, fossil fuels, and water, and what is involved getting them to market, the scale for formation and depletion of natural resources in time and space, and whether sustainability can be achieved.

8. Climate Change

Objective: Students will be able to analyze and explain the Earth's changing climate over various time scales and analyze the environmental, social, and geological impacts of these changes.

The Earth's climate is changing and has throughout geologic history. Students should know what climate change is, the difference between climate and weather, how climate has changed on the geologic time scale, and how it is changing on a present-day time scale. They should be familiar with and know how to use climate proxy records (e.g., ice and sediment cores, tree rings, speleothems, etc.). Students should understand rates of change and how gradual versus rapid change creates different impacts and effects on global, continental, and local scales. Students should examine the driving forces and causal mechanisms of climate change, and how much dependence there is on the spatial and temporal scales of change. Students should also know the difference between external and internal climate forcings, what feedback mechanisms are acting, and how they work. They should consider climate change from an Earth systems perspective, including the impacts of plate tectonics, atmosphere-ocean-cryosphere-Earth interactions, the carbon cycle, the role of oceans, and human-induced climate change. Students need to analyze the impacts of climate change in the context of a variety of important societal issues: impact on water resources and the hydrologic cycle; implications for the biosphere, ocean acidification, and sea level rise; effects on soil and agriculture; impacts on ocean-atmosphere circulation; and the economics and societal aspects of climate change. Students should be aware of the climate element in environmental consulting, in hydrogeology, and in petroleum exploration.

Additionally, participants at the *2015 Geoscience Employers Workshop* thought fundamental physical and chemical processes and concepts, such as thermodynamics, kinetics, diffusion, energy, heat and mass transfer, fluid flow, and key geochemical cycles (e.g., C, H_2O, N, P), should be introduced where applicable to the study of the geoscience concepts discussed above.

B. CORE GEOSCIENCE AND SCIENCE SKILLS FOR SUCCESS

At the *2014 Summit*, participants recognized that while many skills were common to science, others were specific to the geosciences. Most academics and employers in the *2014–2015 survey* identified the following core science and geoscience skills as either important or very important (Figs. 3-2, 3-3).

Skills common to all sciences that students should develop include:

- critical thinking and problem solving;
- communicating effectively to scientists and non-scientists;
- accessing and integrating information from different sources and to continue to learn;
- understanding and using scientific research methods;
- applying strong quantitative skills; and
- working in interdisciplinary teams and across cultures.

Employers at the *2015 Geoscience Employers Workshop* agreed students should be both proficient in these skills and have experience applying them. Employers felt the capability to access and integrate information from different sources while continuing to learn was something students needed to master. The *2014–2015 survey* results showed that a substantial majority of both academics and employers (84–95%) viewed these science skills as important or very important, with the exception of working in interdisciplinary teams and across cultures where a smaller majority (65%) agreed (Fig. 3-2) (Appendix A).

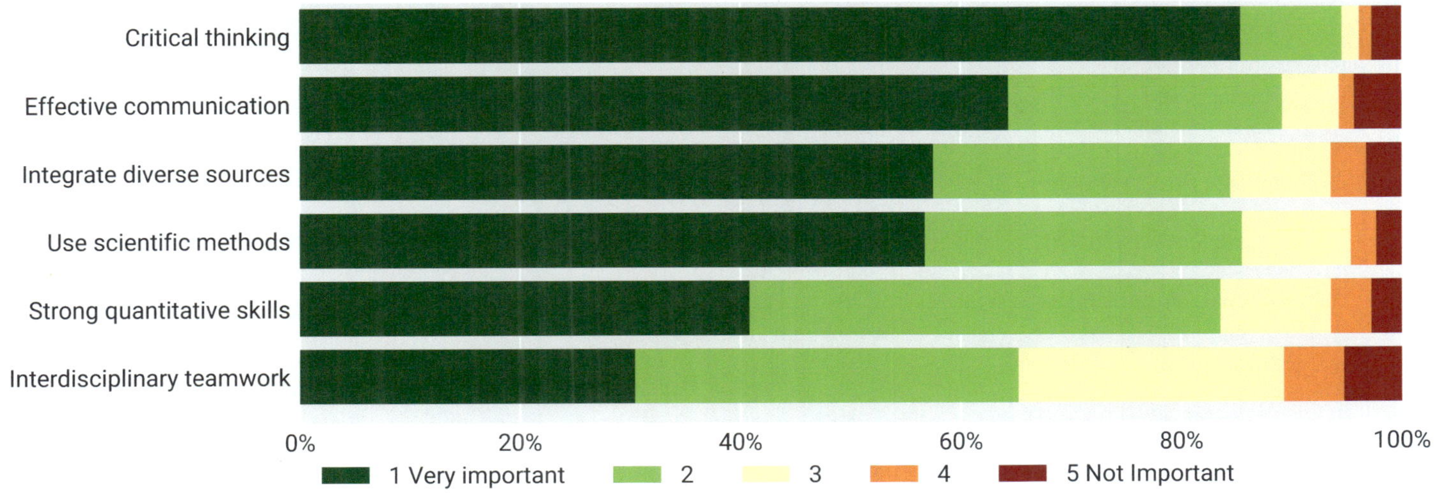

Figure 3-2: Science Skills Importance

Survey Question: The Summit report identifies the following skills as critical to undergraduate education. For each, please indicate the importance from your perspective of these skills for undergraduate-level curricula.

Summit participants and employers also identified several geoscience-centric skills that students should develop (Fig. 3-3). These included:

1. making inferences about the Earth system from observations of the natural world combined with experimentation and modeling;
2. solving problems requiring spatial and temporal (i.e., 3D and 4D) interpretations;
3. working with uncertainty, non-uniqueness, incompleteness, ambiguity, and indirect observations;
4. integrating data from different disciplines and applying systems thinking;
5. applying field observation and integration skills and a working knowledge of GIS;
6. using strong computational skills and the ability to manage and analyze large datasets; and
7. being technologically versatile (i.e., Google Earth®, tablets, smartphones, apps).

Of these seven geoscience skills, 80–87% of *2014–2015 survey* participants saw the first four as either very important or important while 52–67% viewed the last three as very important or important (Fig. 3-3) (Appendix A). By contrast, *2015 Geoscience Employers Workshop* participants felt that students should have a demonstrated mastery of five (#s 1, 2, 3, 5, 7) of these skills by having completed a project or thesis. For the other two (#4 and #6), they expected proficiency through application of these skills in their coursework or projects. Employers at the workshop stressed the importance of all these science and geoscience-specific skills, including those that garnered lower percentages in our survey responses.

C. CORE COMPETENCIES

Both academics and geoscience employers agreed that developing competencies — the ability of students to use the concepts and skills they had learned — are important. At the *2015 Geoscience Employers Workshop*, employers expanded on the skills identified by the academic community, identified what these skills meant in a work environment, and detailed what competencies students need to be successful in the workforce.

Employers expected, for any of the identified science and geoscience skills, at least a level of proficiency — students had experience using or applying what they knew, most likely in class exercises. For geoscience-specific skills, employers generally expected mastery — students had demonstrated the ability to use skills through a project or thesis. Only exposing students to these broad concepts with the development of some understanding was viewed as insufficient. Students must be able to use what they know.

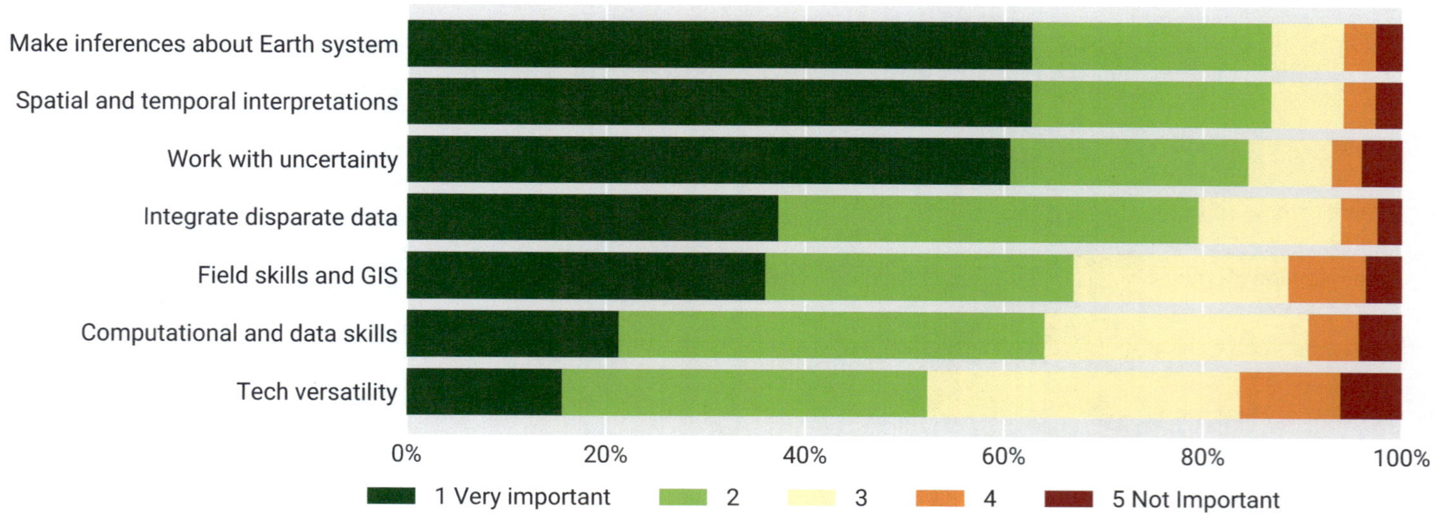

Figure 3-3: Geoscience Skills Importance

Survey Question: The Summit report identifies the following skills as critical to undergraduate education. For each, please indicate the importance from your perspective of these skills for undergraduate-level curricula.

The important competencies for skills, and to a lesser extent for concepts, that geoscience employers highlighted and were endorsed by participants at the *2016 Heads and Chairs Summit*, are discussed in the following subsections. Starting with the adjacent observation from the community, and throughout this document, quotes from the heads and chairs progress reports illustrate how the results of this initiative have been applied (see also Section 12; Appendix C).

Geoscience and Systems Thinking

Employers emphasized that *Geoscience Thinking and Systems Thinking* are core competencies. In this context they meant that *Geoscience Thinking* (or geoscience habits of mind) requires thinking on geologic and real timescales, spatial thinking in 3D (and 4D), and direct field observations, coupled with geologic reasoning and synthesis (Kastens et al., 2009). When students solve geoscience problems, they need to think of Earth as a complex, open, and dynamic system of interlinked parts, processes, and feedbacks. Students need to examine how the atmosphere, hydrosphere, lithosphere, pedosphere (surface), and biosphere work as a system and interact, including the driving forces for change and their effects, and the coupled solar system-Earth interactions. Additionally, students should be able to incorporate the human element — the influence of geosciences on society and of society on Earth processes and its impact in shaping the human experience — the coupled human/societal-Earth system.

Critical Thinking and Problem Solving

Preparation for "real world" professional projects and future research requires strong critical thinking and problem-solving skills. Students need experience solving problems with authentic data and non-unique answers in the context of an open and dynamic system. Employers note that students need to understand the context of the problem and identify the appropriate questions to ask, what data to collect, and methods to use. They also need to be able to collect and compile that data, analyze its quality, interpret it, and apply their results and conclusions. What makes this process distinctive to the geosciences is the need to make predictions with limited data while recognizing and managing uncertainties. Students must have experience working on questions with no clear answers and substantial ambiguity, where they work by analogy and inference, and within the limits of certainty. To solve most geoscience problems, they also need the ability to visualize and address 3D and 4D questions involving spatial scales that may vary from atomic to global, and time scales ranging from geologic to "instantaneous."

Employers are interested in students who have a passion for solving problems and are intellectually flexible enough to apply their skills in new situations. Conducting authentic research and the collection of new information were considered valuable student experiences. Most importantly, students needed to be prepared for and embrace life-long learning. They need to be willing to learn and apply new concepts, ideas and data, use new technology and software; and be able to critically evaluate literature so that as new information becomes available they can assess its validity and usefulness.

Quantitative and Computational Skills

Higher-level math and computer programing skills are increasingly critical for geoscientists, and competency in these

OBSERVATIONS FROM THE COMMUNITY:

We established a new Geology program. Although on paper it is a set of content-specific courses, it represents a collective, conscious shift of focus toward building core student competencies and "must-have" skill sets and attributes in addition to content-driven knowledge base. Placing greater emphasis on students using, applying, communicating, and being innovative with what they've learned, the new program design fills more gaps and contains more purposeful scaffolding. *(Bachelor's-granting public university)* skills have resulted in demonstrated increased employability and employment resiliency. Employers across the spectrum of occupations expressed the importance of these skills. Employers expect students to take calculus and statistics. The latter viewed as the most important as it relates to understanding probability and being able to analyze uncertainty and risk. All geoscience professions rely heavily on these statistical skills as most applied geoscience problems are fundamentally risk assessments.

Additionally, differential equations and linear algebra are highly recommended. Students will need to have a fundamental grasp of differential equations to effectively understand many of the key concepts in fluid flow, which is a pervasive concept in geoscience today. Linear algebra prepares students to understand how complex multivariate systems behave and how multiple variables and dependencies between them can generate multiple solutions. Employers at the *2018 Graduate Student Geoscience Employers Workshop* expected that undergraduates would already have competency in all these mathematical areas. Students wishing to continue for master's and doctoral degrees should be encouraged to start developing competency in higher-level mathematics. Another increasingly important expertise highlighted by employers at both *2015 Undergraduate* and *2018 Graduate Geoscience Employers Workshops* is computational methods, including computer programming and modeling.

Data Analysis Skills

The analysis and management of large datasets is increasing in importance rapidly as a competency, even more now than was evident during the *2014/2016 Undergraduate Summits* and *2015 Undergraduate Geoscience Employers Workshop*. In 2015, employers agreed that students need to be able to examine large datasets and statistically analyze the data to draw conclusions about the information within them; to model data using visual models, modeling tools (e.g., Stella, Modflow, Matlab, etc.), and simulations; and to integrate multiple large datasets of different types and from different disciplines. The increased emphasis on these data skills and competencies, and all levels of data analytics at the *2018 Graduate Student Geoscience Employers Workshop*, was striking. Although these employers were discussing graduate students, the employers made clear that the expectations for undergraduates entering graduate school for these competencies had increased over the three-year period. Overall, employers stressed that successful geoscientists are expected to integrate technical and quantitative skills, programming, and application development. As the geosciences evolve, the need to be technologically facile and diverse, and to have strong data skills, will continue to increase.

Communication Skills

Communicating science verbally and in writing is critical. Employers at *2015 Geoscience Employers Workshops* stressed that students must be able to tailor their written and verbal communication to different audiences, ranging from specialists within their field to other scientists and engineers, educated non-scientists, potential funders, management, the general public, and the press. Another key component of communication stressed by the employers is good listening skills and the ability to carry on an interactive dialogue. They noted that the best leaders know how to listen to ideas and respond constructively. Many heads and chairs in their action plan reports detailed how they embedded communication throughout their curriculum (e.g., Box 3.1).

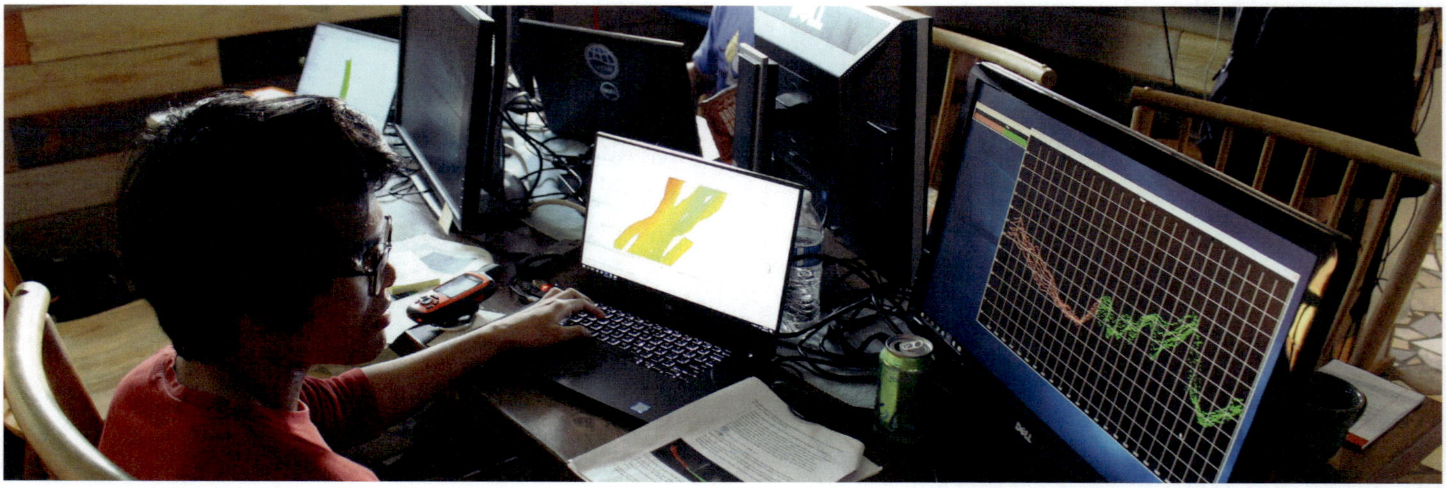

Courtesy of the Jackson School of Geosciences, University of Texas at Austin

Cross-disciplinary Teamwork on Interdisciplinary Projects and Project Management

All employers participating in the workshops stressed the importance of students learning to work in teams, particularly with people from different disciplines, on projects that are commonly multidisciplinary and even transdisciplinary. In the work environment, teams comprise people with different backgrounds, specialties, experiences, and personalities. The team works together on the entire project so that this diversity of thought, experience, and specialties enhances the outcomes.

Project management in a team setting is very important. Each person must learn to be a leader and a follower, to listen and share, and to work with people who have different opinions and approaches. Teamwork requires goal setting and solution-oriented approaches, as well as understanding effective methods for addressing conflict resolution. Strategies for identifying and resolving problems as they arise are essential for projects to succeed. Students need to learn time management to assure that projects are completed in the time allotted. Effective team members need to learn to monitor their own behavior and to set aside interpersonal issues with peers, supervisors, and employees. Other important interpersonal skills are the ability to work with different personalities, emotional makeups, viewpoints, specialties, educational backgrounds, and abilities, including with people one does not like.

Field Skills

Employers strongly supported development of field skills, particularly through intensive experiences such as field camps, research projects, expeditions, and cruises.

Box 3.1: Embedding Skills Throughout Curriculum

Science communication is intentionally embedded into key parts of the curriculum, helping students increase their capacity for effective written, oral, and graphic communication to both scientific and non-specialist audiences. (Bachelor's-granting private college)

A few examples:

- A place-based geology project in the introductory course connects students' prior experience with their novice-level understanding of geologic processes and products. Students write throughout the semester to produce an observation-oriented geologic interpretation of a particular place that is familiar to them.
- Evolution of the Earth (historical geology) includes both scientific writing, with a focus on outcrop- and hand sample-scale description as the basis for geologic interpretation, and public communication, in the form of museum exhibits.
- Earth Surface Processes (geomorphology), which has always included an individual research component, now has an increased emphasis on communicating those findings to both scientific and non-scientific arguments. Students are challenged to think about real-world application alongside their scientific inquiry.
- Research in Geology I and II, our junior-senior seminar course, in which students propose, conduct, and write a senior thesis, has an increasing emphasis on communicating to a non-specialist audience. While students continue to write a scientific paper-style thesis, they must also give a 3-minute 'elevator' speech to a non-specialist audience. We asked non-geology faculty to stop our students in the hallway and inquire about their work. Geology students participate in an all-campus research presentation day, in addition to the scientific research symposium that we have always done. In addition, we explicitly discuss the value of liberal arts education in shaping who they are as scientists and help them craft a personal narrative that can form the basis of a cover letter, personal statement, or interview response.

OBSERVATIONS FROM THE COMMUNITY:

Pennsylvania: Students who graduate without field camp cannot be certified as a Geologist in Training until educational requirements are met. The formal education required must include field geology and structural geology coursework that is sufficient to demonstrate that the candidate has educational experience in tectonics and fractured bedrock geology and the field methods needed to measure, map, and evaluate geologic data.

Fieldwork improves spatial cognition and creative problem solving and provides an excellent opportunity for students to gain experience synthesizing different types of geoscientific information to solve problems. Depending on the design, these experiences can also help students learn teamwork. The employers considered field experiences as unique, essential, and difficult to replicate or substitute. In some states (e.g., Pennsylvania, California) licensing for geologists requires field courses.

Essential Non-technical Skills

Many skills stressed by geoscience employers are not technical in nature but are nevertheless necessary for success. Ethical behavior and adhering to codes of conduct for professions, institutions, and employers is critical for trustworthy scientific results that guide decision-making. Also, awareness of implicit biases and the components of an inclusive environment promotes a productive workplace.

Other non-technical skills or competencies that are not specific to the geosciences but are key to success in the workforce include professionalism, demonstrating a commitment to doing an effective job, being responsible, dependable, honest, confident, committed to effective performance, time management, and generally having a professional appearance. Leadership involves effectively guiding others to accomplish goals or objectives in a coherent and cohesive manner. Business acumen, or some knowledge of the business environment that provides the ability to make good business decisions, is very beneficial. Risk management is critically important: most geoscience careers involve making decisions that include financial, environmental, structural, or other types of risks.

Geoscience is global. Not only is the science and impact global, the range of geoscience employment and colleagues are global in scale. Geoscientists must have a global perspective and the ability to work with people from different cultures. Additionally, geoscientists work on issues of importance to society, and employment in these areas is increasing rapidly. Thus, students need to develop an understanding of the societal relevance, as stressed in many of the concepts outlined above, and the ethical implications of their work.

APPROACHES TO DEVELOPING CONCEPTS, SKILLS, AND COMPETENCIES

Employers at the *2015 Geoscience Employers Workshop* and Heads and Chairs at the *2016 Summit* discussed the best ways to develop skills, understanding of concepts, and competencies. Experiential learning was considered instrumental to achieving identified student learning outcomes, either within existing classes or as distinct experiences. Students should be constantly engaged in practicing their skills and using the concepts within classes. Gaining experience in problem solving using and analyzing real datasets is critical. Students also will benefit by using methods and geoscience equipment and tools to gather data and solve problems. Written and oral work should be integrated into classes, and intensive writing and oral presentation courses should be part of programs. Students should have opportunities to work on collaborative, integrative, and interdisciplinary team projects. Also important is interactive use of technology to gain experience with visualization, simulation, and modeling of real data.

In addition to experiential learning in existing classes, substantial experiences such as fieldwork and field experiences, and capstone problem or project-oriented courses, are valuable. Independent research experiences and projects, senior theses, and internships or REUs (Research Experiences for Undergraduates) provide excellent ways to develop skills and use concepts. Employers also stressed the importance of active collaboration between academia and outside employers (discussed in Section 13).

A major concern expressed at the *2016 Summit* and subsequent workshops was that most geoscience undergraduates were not prepared for the math, quantitative skills, data analysis, and computational skills recommended by the employers. At the undergraduate level, only about 25% of the graduates, and about half of advanced degree recipients, have taken substantial math beyond Calculus II (Wilson, 2018). Geoscience employers across all employment sectors, including atmospheric and ocean sciences, highlight quantitative geoscience skills as critical for bachelor's graduates. The rapid growth of data-intensive applications in both basic and applied geoscience investigations across subdisciplines highlights the need for more advanced experiences with statistics. The ability for students to manage data and use numeric data systematically with a full understanding of the limits of accuracy and precision, and appropriate use of transformative functions and algorithms, is now central to the nature of geoscience work.

Heads and Chairs at the *2016 Summit* discussed ways of developing these quantitative, data analysis, and computational skills, ranging from increasing external course requirements to integrating these skills into geoscience courses, and some cite examples of success in their progress reports (see Box 3.2; Section 12). As much of the mathematical and statistical education of undergraduates happens outside of geoscience departments and programs, it is critical that math and other quantitative skills be integrated into geoscience courses throughout the curriculum. Too often, math is not used in geoscience courses, even if it is a prerequisite or required for graduation, although the situation is improving (McFadden et al., 2019). Faculty have tried a range of strategies to ensure the quantitative abilities of their graduates, from specialized courses on mathematical applications in geology (e.g., Vacher, 2000; Ricchezza and Vacher, 2017), to developing and leveraging a range of online resources in courses to support students' quantitative skills. The NSF-supported "Math You Need, When You Need It" project (Wenner and Baer, 2015; Wenner et al., 2011) produced a set of online resources aimed at supporting student understanding of common mathematical applications encountered in introductory-level geoscience courses, such as linear regression, making and reading graphs, and understanding rates, among others.[1] One specific example for both introductory and upper-level geoscience and other STEM curricula is the Spreadsheets Across the Curriculum educational resource collection[2] (e.g., Vacher and Lardner, 2010) that leverage the computational functions in Microsoft Excel to help students learn different aspects of mathematics. Heads and chairs in their progress reports cited ways they were increasing the quantitative rigor of their programs (e.g., Box 3.2)

©Shutterstock/cherries

1 See https://serc.carleton.edu/mathyouneed for the full collection.

2 https://serc.carleton.edu/sp/ssac

Box 3.2: Increasing Quantitative Rigor of Undergraduate Programs

Many departments have started increasing the rigor of their undergraduate programs.

One Bachelor's-granting private university came up with several different approaches to integrating more math and computational and data analytics skills:

- All elective courses focus on developing skills and conceptual understandings valued by industry, including quantitative, written, conceptual, and technical skills.
- We led an effort with the business and math departments to propose a new cross-disciplinary B.S. degree in GeoBusiness and Data Analytics. Students take core geology classes and courses in business, economics, statistics, big data management, spatial data analysis, programming, and technical writing. The integrated skill set developed was viewed by our advisory board and other industry experts as very strong and employable. The Academic Vice President saw it as an innovative degree that will prepare students for the business workforce and for non-geology graduate work (MBA, geo-economics, etc.). He is encouraging other departments to consider similar cross-disciplinary degrees.
- Many of our students have an interest in geospatial studies so we created a B.S. degree that builds on that interest including core geology classes as well as and earned certificate in GIS and in another technical field including computer programming, web design, networking, or data science. Graduates with this degree have found immediate employment.

Other doctoral granting institutions have worked to overcome limited quantitative and computer skills among their undergraduates by introducing more quantitative activities across a range of courses and including examples in every course to show how these skills are important and applied in Geological Sciences. Several have started offering their own quantitative or computational geoscience courses.

Geospatial information science (GIS) skills were also highlighted as a needed addition to geoscience programs during both the *2014 Summit* and *2015 Geoscience Employers Workshop*. Since then, the importance of student statistical skills related to dealing with large (and geospatially controlled) Earth datasets has increased. Geospatial skills and reasoning, including spatial statistics and analysis, are highly marketable and are in high demand by students. These have been incorporated into some bachelor's degree geoscience programs, though others recommend that students seek coursework in this area through their general electives.

Summit and workshop discussions about GIS, quantitative/computational skills, and other competencies that traditionally have been addressed through supporting science and mathematics courses centered on the lack of alignment of these offerings with the context and application needs of the geosciences. Some programs have moved to models where formal supporting courses are paired with a geoscience-focused offering in the same disciplinary area (e.g., a course in physics with a follow-on geophysics course; a course in chemistry with a follow-on geochemistry offering, etc.) to give students more context with these allied sciences.

With GIS, and in some cases quantitative/computational content, certificate programs offered at two- and four-year institutions, and/or badging, provide a mechanism for students to document these experiences and capabilities on resumes and in graduate applications. Some programs are seeking to leverage the credential trend, encountered more often in master's programs, of stackable certificates in geoscience subfields, several of which can be compiled by students to "construct" their bachelor's degree.

Heads and Chairs at the *2016 Summit* expressed concerns that if the requirements for a geoscience degree were too rigorous, they would have problems attracting and retaining students to their majors. Feedback from the geoscience employers, however, stressed that this was doing graduates a disservice as the quantitative skills were needed for employment and the level of skills needed was increasing over time. Counter to the concern of enhanced quantitative requirements being a dissuasive factor for majors as expressed by academic participants, it is important to recognize that strong quantitative skills are a predictor of employment resilience (Keane and Wilson, 2018). Workers who completed courses in differential equations, linear algebra, and computational methods have not been displaced as quickly as individuals who lack these skills. One issue for consideration in their success, beyond the motivation for a higher level of preparation, is the cognitive development conveyed through these courses. As noted above, differential equations and linear algebra better prepare students to understand many key concepts in fluid flow and complex, multivariate systems. This cognitive development is extremely beneficial to developing a mastery of broad geologic thinking. Additionally, computational methods courses represent a core tool in appropriate data modeling and management.

Recommendations:

- Provide students through their courses and activities the opportunity to develop an understanding of broad concepts, including processes and impacts, to build a working framework for knowledge gained during their education and future career
- Incorporate instruction and practice in geoscience skills identified by employers and academics across multiple classes to ensure students gain sufficient competency in these skills
- Provide students with authentic experiences that incorporate geoscience and systems thinking and problem solving (e.g., field experiences, research projects, in class exercises with real data, etc.)
- Incorporate the development of key professional skills (communication to diverse audiences, teamwork, project management, etc.) and those skills more closely aligned with the discipline into your undergraduate program
- Identify the key quantitative reasoning skills required for your graduates and incorporate practice in these skills at multiple points in their degree program
- Provide students experience and practice in acquiring and analyzing real data using multiple methods and tools to solve geoscience problems (if practical for your institution) and handling large data sets, with full understanding of accuracy, limitations, and uncertainty

Courtesy of the Jackson School of Geosciences, University of Texas at Austin

Courtesy of the Jackson School of Geosciences, University of Texas at Austin

4. Approaches to Implementing Curriculum Reform: Backwards Design and "Matrix" Strategies

The community vetted concepts, competencies, and skills provides the basis for successful curriculum revision in which student learning outcomes become the foundation of curricula planning.

A major conclusion of the 2014 and 2016 *Summits, 2015 Geoscience Employers Workshop and 2014–2015 survey* is that developing competencies, skills, and conceptual understanding is more important than taking specific courses. Faculty and other geoscientists commonly find it difficult to agree on what specific courses students should take, however they generally agree on what students should learn and be able to do. These student learning outcomes — what *students* should know, be able to do, and demonstrate when they have completed a course or program — should be the basis for any curricular revision. Many successful curricular revision efforts in the geosciences have worked from the "backwards design" principles outlined in Wiggins and McTighe (2005; see also McTighe and Wiggins, 2012), in which student learning outcomes, as the desired goals, become the foundation of curricular planning. The community vision for concepts, skills, and competencies, described above, provides the basis for developing bachelor's geoscience curricula and programs.

The first step towards significant curricular revision is establishing departmental consensus on curricular learning outcomes, followed by defining how well these outcomes are being met by the current program. The keys to success are:

1. attain faculty agreement on concepts, skills, and competencies that their undergraduate students should develop;

2. carefully analyze the current curriculum and/or extracurricular activities to discover whether the course sequence builds these core elements and to find gaps and unnecessary redundancies; and

3. redesign curriculum including course content and sequence to meet agreed upon student learning outcomes.

Listing concepts, skills, and competencies along one axis of a matrix, and the current courses along the other, allows individual faculty to indicate which of these they cover or develop in their class and to what extent or depth. The matrix then forms the basis for making changes to the curriculum. Examples and helpful information can be found at the National Association of Geoscience Teachers' (NAGT) Building Strong Departments website under Design Degree Programs (https://serc.carleton.edu/departments/degree_programs/matrix.html). A modified approach is to focus on the big picture reforms, concentrating on building a new curriculum based on the agreed upon student learning outcomes, and doing a course-by-course matrix later.

OBSERVATIONS FROM THE COMMUNITY:

"Developing the online matrix of competencies/skills in order identify those common to multiple courses was very helpful." The faculty met to identify and construct a matrix of undergraduate course competencies/skills. The matrix was posted online and faculty was asked to identify which competencies/skills are taught in their courses. *(Doctoral-granting public university)*

Use the concepts and skills matrix to your advantage, as an instrument that was nationally vetted by geoscience faculty and employers. Sharing summaries from the Summits, along with the employer-vetted concept and skills matrix, helped greatly with faculty "buy-in." *(Master's-granting public university)*

OBSERVATIONS FROM THE COMMUNITY:

"Many/most of the faculty were ready to buy into the curricular changes. We learned from an NAGT site visit (the Building Strong Geoscience Departments program) how to develop and implement a good course and curriculum assessment plan." *(Master's-granting public university)*

We hosted an NAGT Traveling Workshop called "Building Stronger Geoscience Departments" where we created an action plan for curricular revision and followed through with numerous department meetings dedicated to developing a new curricular model, including sharing it with a student focus group for feedback. *(Master's-granting private college)*

OBSERVATIONS FROM THE COMMUNITY:

After implementing a matrix approach, we have a more cohesive curriculum that stresses repeated exposure, expected mastery of key skills necessary for conducting research, reporting research (oral and written), analyzing data, and designing a research plan. *(Doctoral-granting Hispanic Serving R2 public university)*

A well-documented example of the Backwards Design and matrix strategy comes from David Mogk (Mogk, 2016), who used it for gathering information about depth of coverage of key skills and competencies in the Geology and Geography degree programs at Montana State University (see also Mogk, 2013, 2014, 2015; Savina et al., 2001). There, the faculty identified key discipline-based and transferrable skills and competencies central to bachelor's graduates. Data on the coverage of these key competencies, concepts, and skills in their majors and supporting courses were then collected from the instructional faculty, based on their syllabi and course activities, and vetted by the Chair. These data were mapped by course and by faculty member in a matrix. This compilation was then used to identify gaps in coverage (i.e., important concepts/competencies/skills not currently covered) and depth of coverage (e.g., to what extent were key skills/competencies revisited and reinforced through the program, assuming at least three opportunities for practice were necessary to develop mastery; Mogk, 2016). This effort informed faculty decisions regarding the revision of courses and/or course content and activities, and the revised matrix became the foundation for establishing and measuring student learning outcomes required for University accreditation (Mogk, 2015).

Designing a curriculum focused on student learning outcomes makes assessment simpler. The course-by-course curriculum matrix serves as a blueprint of expected student learning outcomes, which can be used for any university or department-wide assessments or accreditation. Further, faculty know more about what students should have learned and done in prior courses and can build on these concepts, skills, and competencies.

A developed matrix should be shared with students so they can see how they are progressing across their program of study, recognize if they have developed the skills necessary to be successful, and help them identify areas where they may want to supplement the curriculum with other activities.

About 50% of progress reports from participating heads and chairs at the *2016 Summit* and subsequent workshops stated that they used a matrix approach and the community vetted concepts, competencies, and skills for bachelor's programs as a guide, but with modifications based on the type and size of the program. A common result was faculty went into the exercise convinced everything was fine with the curriculum and came out surprised by the large number of gaps and unnecessary redundancies. Most of the departments used the matrix approach to describe the current program and design a set of recommendations for curriculum reform. A few skipped analyzing the current program. Changes included modifying existing course content and designing new courses that merged content from others courses or introduced new content. Importantly, this effort resulted in specifically embedding skills into courses and developing a sequence so students had repeated opportunities to develop and attain a mastery of key skills. Course sequences were revised along the lines of the core-competency goals and implementation of competency-based bachelor's curricula.

Faculty retreats that focused on curricular redesign were found to be extremely useful by departments across the range of institutions. Some brought in NAGT's Traveling Workshop "Building Stronger Geoscience Departments" to campus and others found resources at the Science Education Resource Center (SERC) at Carleton College (serc.carleton.edu) on retreat planning and Backward

Curriculum Design helpful. Sharing of course syllabi and/or learning outcome goals (i.e., concepts, skills) in prerequisite and core courses helped in addition to using backwards design.

Recommendations:

- Establish a geoscience faculty and/or department consensus on curricular/student learning outcomes using recommended community vetted concepts, skills, and competencies, accounting for institutional priorities and capacities
- Analyze how well these outcomes are being met by current curriculum and courses
- Redesign curriculum including course content and sequence to meet agreed upon curricular/student learning outcomes

OBSERVATIONS FROM THE COMMUNITY:

"When we fully developed matrices for all three of our undergraduate degree programs, we discovered we had placed much of the transferrable skills in specific classes (e.g., writing, in one course titled Scientific Communication) which meant students would get these skills once, but maybe not again. Alpha-testing infused these transferable skills into the majors courses (e.g., writing exercises, taught by a second instructor who specializes in scientific writing, embedded in the context of an exercise in Structural Geology)." *(Master's-granting private university)*

"Working within the structure and course requirements of the existing degree path (B.S. in General Geology), we broke our existing outcomes into more granulated outcomes, added competencies and skills (where needed), and mapped these courses into a Mogk-type matrix. We collected syllabi and materials and met individually with all faculty to verify which outcomes and skills were met by the course and at what level (basic, enriched, reinforced). We have assessed our outcomes (one course per year) via mandated degree assessment from our institution starting with our introductory course and moving up through the curriculum. *(Doctoral-granting R1 public university)*

Faculty buy-in may be facilitated by progressing slowly enough to allow feedback. For example, for two weeks the department filled two walls with posters listing their learning goals. This allowed everyone an opportunity to provide feedback. *(Doctoral-granting R1 public university)*

Tim Gibson for AGI's 2018 Life as a Geoscientist contest

5. Best Practices for Instruction of Geoscience Undergraduates

Geoscience educators should further embrace active teaching strategies that research has shown to improve student learning.

Courtesy of the Jackson School of Geosciences, University of Texas at Austin

In the last few decades, new approaches to geoscience instruction have developed based on the results of discipline-based education research (Tewksbury et al., 2013; McConnell, 2019). These new teaching strategies support active learning and build on the traditional mix of classroom, lab, and field instruction to foster deep conceptual understanding of important skills and competencies (e.g., Manduca and Mogk, 2006). Many participants at the *2014 Summit* were unaware of these pedagogies and the research supporting their usage. The participants concluded that the community's primary pedagogical challenges lie in encouraging wider awareness and adoption of these practices and in characterizing their educational impacts and benefit.

Over the subsequent years, community discussions and many NSF-funded geoscience education projects, including this one, have increased awareness and adoption of research-supported pedagogies. In the *Summit 2014–2015 survey*, 79% of the 360 responding academics said their department was interested in making changes in how they did undergraduate teaching; 46% indicated their department was already making, or about to make, systematic efforts to encourage faculty to incorporate research-validated teaching strategies. Also, 40–53% of respondents thought that all or most of their program's faculty were instituting some of these strategies, but that less-common strategies were used less frequently (2–22%) (Fig. 5-1). The results of the *2016 National Geoscience Faculty Survey* (Egger et al., 2019) show a deeper penetration of these methods across departments with the proportion of geoscience faculty incorporating active learning instructional strategies in their courses climbing from one-third in 2004 to over half in 2016. This corresponds to increases in the frequency of use of small-group discussions and in-class exercises (Manduca et al., 2017; Egger, 2019; Egger et al., 2019; McFadden et al., 2019).

Generous NSF-DUE funding over the last twenty years has supported professional development opportunities for geoscience faculty, many through the long-running On the Cutting Edge project (MacDonald, et al., 2004), and have generated large collections of high-quality geoscience teaching resources, along with detailed "how to" explanations for a range of research-validated instructional strategies. The National Association of Geoscience Teachers (NAGT), working with the Science Education Resource Center at Carleton College (SERC; serc.carleton.edu), stewards these collections, providing geoscience instructors with free access to a rich library of shared resources that can support a transition to research-validated teaching strategies (Manduca et al., 2010; McConnell, 2019; Box 5.1).

Box 5.1: SERC and the NAGT On the Cutting Edge Program

The 17-year NSF-funded On the Cutting Edge project (2002–2019) offered a broad menu of professional development workshops for faculty[3] and produced, aggregated, and peer-reviewed an extensive menu of teaching resources, which are available to geoscience faculty through the Teach the Earth collections at the Science Education Resource Center (SERC) at Carleton College. In 2019, On the Cutting Edge transitioned from NSF support to becoming an initiative jointly managed by the National Association of Geoscience Teachers (NAGT) and SERC. The NAGT On the Cutting Edge professional development program supports building capacity for instructional reform through the combination of professional development experiences and online materials. These programs disseminate effective curricula and pedagogical resources, provide opportunities for instructors to reflect on their teaching and learn from trusted colleagues, and promote the development of a shared vision within a community of practice (Kastens & Manduca, 2017).

The Teach the Earth portal[4] provides access to online resources related to teaching and learning, and programs such as Pedagogy in Action[5] are focused on describing research-based instructional strategies and explaining how they can be adapted and implemented in geoscience courses.

Major geoscience professional organizations (American Geoscience Institute (AGI), The American Geophysical Union (AGU), the Geological Society of America (GSA), and NAGT) have collaborated to create the Traveling Workshops Program[6], which is intended to help guide faculty and departments through critical disciplinary and institutional changes.[7]

The success of these initiatives has highlighted the need for sustainable professional development experiences for geoscience instructors. To meet this need, the National Association of Geoscience Teachers established the Earth Educators' Rendezvous[8] as a stand-alone, five-day professional development event, offering a mix of programming that includes multi-day workshops, mini-workshops, research presentations, round table discussions, and plenary sessions. The Rendezvous attracts more than 300 geoscience educators each summer and is an accessible forum for instructors to get a crash course in those pedagogical components that will help them to reform their coursework.

3 https://serc.carleton.edu/NAGTWorkshops/about/workshops.html

4 https://serc.carleton.edu/teachearth/

5 https://serc.carleton.edu/sp/

6 https://www.nagt.org/nagt/profdev/twp/

7 See full listing of the NAGT On the Cutting Edge services and support at https://nagt.org/nagt/profdev/oce.html

8 https://serc.carleton.edu/earth_rendezvous/

About 40% of the progress reports from participating heads and chairs at the *2016 Summit* said usage of active and experiential learning increased. The most impactful action that increased usage was professional development for faculty, including attending workshops (e.g., Earth Educators Rendezvous). Some heads and chairs incorporated the use of active learning teaching into their annual evaluation process, incentivized adopting these methods, or implemented new peer teaching evaluations. Others pointed to incentivization by having enthusiastic junior faculty, peer mentoring, open faculty discussion and sharing of resources, positive outcomes supported by student assessment, and introduction of new ideas at retreats. Some recognized that changing how they taught would help student recruitment and retention.

Continued progress require faculty time to implement active learning and other pedagogical innovations, such as non-instructional assignments, redistribution of workload, or release time and professional development opportunities to learn effective ways to improve student learning. Nonetheless, many of the progress reports cited resistance from faculty to changing how they taught, and many participants in subsequent workshops remain unfamiliar with most active learning pedagogies.

In the following subsections, we discuss these best practices with the goal of encouraging wider awareness and adoption of these research-validated practices and how educational impacts and benefits demonstrate the importance of integrating active and experiential learning into undergraduate programs. Faculty buy-in can be increased by using the concept of scientific teaching (Handelsmon et al., 2007), where faculty rely on evidence that shows what works in teaching.

ACTIVE LEARNING AND BUILDING CONTENT KNOWLEDGE

Active learning in the classroom builds a deeper understanding of geoscience concepts. Students learn better when they can actively monitor their understanding through a variety of activities *during class* (Pollock & Finkelstein, 2008; Derting & Ebert-May, 2010; Freeman et al., 2011, 2014). Adopting these empirically validated instructional practices also increases student retention rates (Russell et al., 2007; Graham et al., 2013) and reduced achievement gaps among different student populations (Haak et al., 2011; Eddy & Hogan, 2014).

Active learning environments typically include two or more of the following elements:

1. Students observing or participating in activities in addition to, or instead of, listening to direct instruction;
2. Opportunities for student reflection on their learning, or student/instructor interactions that lead to assessment of learning;
3. Peer-to-peer interaction among students as they complete activities (McConnell et al., 2017).

Supplementing or replacing lectures with in-class exercises and small group or entire class discussions allows students to think about and use lecture information and to learn collaboratively. In the geosciences, courses with a high level of active learning typically include exercises with the interpretation of maps, graphs, and/or natural datasets (Teasdale et al., 2017).

Many research-validated methods for engaging students in active learning are widely used in both large and small-enrollment courses. (McConnell et al., 2017) discuss the instructional utility and learning efficacy of several of these strategies (discussed below) and summarizes the supporting research. Many of these methods have resulted in improved conceptual learning and retention of specific knowledge, and most have high utility—they are relatively straightforward to prepare and use in almost any class setting. Many of these same methods have had positive impacts on learning, supported by multiple studies across several settings. Instructors who revise a geoscience course may prefer to start by incorporating some of these higher utility strategies as there are many readily available geoscience-specific examples.

The three easiest to use strategies are Think–Pair–Share, minute papers (Box 5.2), and peer instruction (Box 5.3), all of which involve students responding to questions in class and are useful in classes of any size. Think-Pair-Share is commonly used to generate class participation and social interaction between students and can be used as retrieval practice. Minute papers provide a good way to check and correct understanding during lecture.

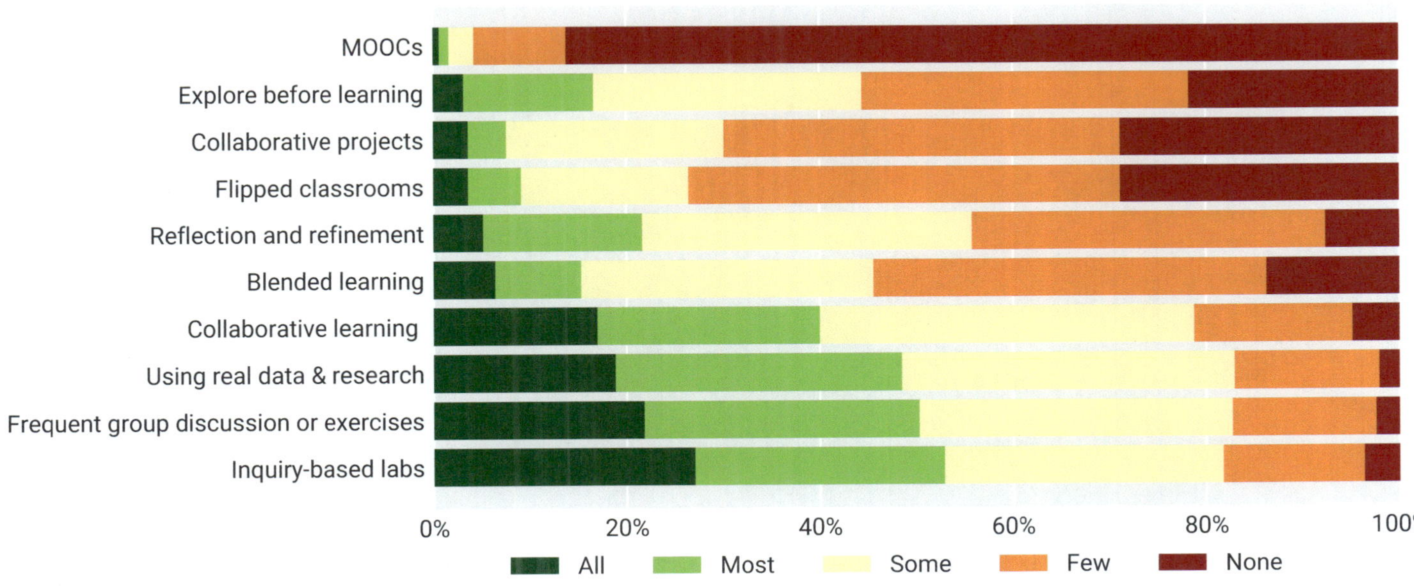

Figure 5-1: Teaching Method Use

Level of use by faculty within a department

Survey Question: From what you know about your department, which of the below teaching methods are used by faculty in your department?

Box 5.2: Easy-to-use Active Learning Practices

Think-Pair-Share[9]. Students are presented with a question and asked to think about their answer, discuss it with a neighbor, and then engage in a full-class discussion. This exercise is particularly useful in large classes and increases constructive social interaction among students. Students may feel more comfortable asking questions and participating in class discussions once they know others have the same questions or understanding of the material. Depending on what is asked, it can also serve as a retrieval practice, where students are asked to recall and use something they learned previously. Although Think-Pair-Share is easy to use, most research studies that have included Think-Pair-Share have done so in combination with other active learning strategies, and there are limited research on the effectiveness of this strategy used alone.

Minute papers[10]. Students are asked to briefly write about what they thought was the most important thing covered during the lecture or what was the most confusing. To be effective, the instructor needs to use the results in the next class period to confirm what was important or clarify what students found confusing. Minutes papers are generally straightforward to implement and are moderately effective at improving learning.

9 https://serc.carleton.edu/introgeo/interactive/tpshare.html

10 https://serc.carleton.edu/resources/14315.html

Peer instruction strategies are one of the most widely adopted active-learning approaches (Crouch and Mazur, 2001), and are well suited to assess students' comprehension of concepts, or their application of concepts to new situations. In the geosciences, ConcepTests (McConnell et al., 2006) are available through SERC and can be readily incorporated into most class settings. Lecture Tutorials are also available but involve more upfront preparation (see Box 5.3). These approaches allow students to assess their learning, identify the concepts and content that are important for the course, and develop communication and reasoning skills among their peers. These strategies help instructors quickly identify the content that most students understand or topics that present a challenge and allow instructors to adjust in-class activities accordingly. Peer instruction has high-to-moderate utility, depending on the specific approach, and its effectiveness in improving student learning is well documented[11].

11 https://serc.carleton.edu/NAGTWorkshops/teaching_methods/conceptests/

Courtesy of the Jackson School of Geosciences, University of Texas at Austin

Other active learning strategies often involve more effort and preparation but are recognized as very effective at improving learning. These include teaching with models, constructing concept maps, case studies, and problem-based learning activities. The common practice of physical models being used during lecture demonstrations, and/or by students in lecture or lab settings, has been shown by research to engage students and improve learning. Concept maps, where students graphically represent what they know about a topic and show similarities among various ideas and concepts, also has been shown to improve learning[13].

Case studies[14] and problem-based learning exercises[15] have the potential to examine real-world science and emphasize application and problem-solving rather than memorization. Such activities permit instructors to engage students in higher-order learning and to address problems that involve the analysis and/or synthesis of large datasets. Using real data, where students work on problems with no exact answers, either in class or in lab, is important for developing problem-solving skills and increasing conceptual understanding. Such exercises may involve case studies that model a specific situation, requiring an action in which students take on different roles reflecting a real scenario (e.g., representing a company, a member of the community, a geoscientist, etc.), encouraging the development of communication and non-technical skills as well as content-specific understanding. These activities commonly involve

13 For Concept Maps: https://serc.carleton.edu/NAGTWorkshops/assess/conceptmaps.html; also see McConnell et al., 2017.

14 For case studies: https://serc.carleton.edu/sp/library/cases/index.html; https://serc.carleton.edu/NAGTWorkshops/undergraduate_research/case_studies.html

15 For problem based learning: https://serc.carleton.edu/NAGTWorkshops/problem_solve/index.html

Box 5.3: Peer Instruction and Lecture Tutorials

Peer instruction. These activities are built around lectures and divided into segments, punctuated by multiple-choice questions (ConcepTests) that focus on key concepts. The students "vote" for the correct answers individually, via clickers, a learning response system such as Learning Catalytics, or a show of hands. If a high percentage choose the correct answer, the class moves on. However, if the question draws a mixed response (i.e., correct responses between 35% and 70%), students are directed to discuss the reasoning behind their choices with neighbors or assigned group members, and then they vote again based on the results of that conversation (Mazur, 1997). Typically, the instructor then discusses the explanation for the correct or best answer before moving on to the next topic.

Lecture Tutorials. The instructor delivers a short lecture on a concept, followed by students working together on a set of relevant questions with the instructor and teaching assistants circulating through the class to provide guidance and support to the individual groups. Group work is followed by an instructor debrief explaining the best answers and answering questions.[12]

12 https://serc.carleton.edu/sp/library/lecture_tutorials/index.html

Courtesy of the Jackson School of Geosciences, University of Texas at Austin

OBSERVATIONS FROM THE COMMUNITY:
"Our focus on streamlining student learning outcomes and methods of assessment has resulted mainly in attempts to integrate more critical thinking, writing, presentation skills, and teamwork into existing classes. There has also been a concerted effort to encourage faculty to engage classes with more active/experiential learning exercises." *(Doctoral-granting R2 public university)*

students working together in small groups on questions while the instructor and/or teaching assistants provide guidance. Again, the activities are followed by a debrief as the instructor solicits ideas from student teams before revealing correct or (in the case of real-world scenarios) historically accurate explanations as to how the issues were addressed and answering student questions.

Case-based or problem-based resources are rarely presented in textbooks, but a growing collection of online resources, hosted at SERC, emphasize the relationship between the geosciences and society, and provide opportunities for instructors to use activities that incorporate real world data[16]. Additionally, some employers are willing to provide datasets and real-world problems for classroom use.

ACTIVE LEARNING IN DEVELOPING SKILLS AND COMPETENCIES

OBSERVATIONS FROM THE COMMUNITY:
Pairing courses throughout the undergraduate trajectory is effective because students see the connections between courses and can spend more time as teams working on projects that span several courses. *(Doctoral-granting Hispanic Serving R2 public university)*

Active learning strategies also develop key skills and competencies, but unlike geoscience content that may be specific to an individual course, skills need to be revisited in multiple classes so students can practice, develop competency, establish mastery, and recognize how these proficiencies are employed. Geoscience instructors should work toward having learning outcomes that target broadly applicable skills and competencies in data analysis, scientific communication, quantitative reasoning, and geoscience and systems thinking.

Scientific communication skills develop only through repeated practice with different types of professional written and oral presentations. Scientific abstracts, "one-pagers" for non-geoscientists, and short grant proposals provide students with a variety of written formats targeting different audiences, while keeping the evaluation and review process manageable for the instructor. Multiple short assignments involving review and editing in several courses provide more useful communication practice to students than term papers. Similarly, short student presentations in classes targeted to different audiences provide students with chances to improve their presentation skills. Although technical writing classes can be effective, it is important to integrate these skills throughout the geoscience curriculum.

Geoscience thinking involves spatial thinking, particularly visualizing in three dimensions from field observations, temporal reasoning, and understanding Earth as a complex system (Kastens et al., 2009). In both introductory and majors' courses, students need practice solving 3D spatial and geologic time problems. Students need to be able to describe a natural system in detail, discuss changes that have multiple effects, and recognize and analyze feedback loops. All geoscience students need experience making field observations, conducting field-based investigations, and working with geospatial data.

In the *2016 National Geoscience Faculty Survey*, geoscience faculty reported that approximately two-thirds of their students work with geospatial data at least once during their courses, and more than half indicated that students made field observations (Egger, 2019). Similarly, most instructors, whether teaching introductory or majors' courses, reported that their students practiced both 3D spatial

16 For example, InTeGrate https://serc.carleton.edu/integrate/; GETSI, https://serc.carleton.edu/getsi/.

thinking and temporal thinking at least once, with this kind of practice occurring more commonly in majors' courses. Likewise, a majority of instructors indicated that they featured some aspects of system thinking (e.g., describing a system, discussing a change with multiple impacts) in their courses, while other features of systems (e.g., analyzing feedback loops) were typically not covered.

Although a large proportion of geoscience instructors self-identify as teaching with a focus on a range of skills and competencies (Viskupic et al., 2020), the evidence is less clear as to whether the necessary learning outcomes are successfully achieved. Assessing content knowledge is easier than measuring the development of specific skills and competencies which necessarily develop over multiple courses throughout a curriculum across several years. Community college (2YC) faculty can play an important leadership role in developing best practices for geoscience education. 2YC faculty have a greater capacity to trial-run and enhance active learning pedagogies given the number of opportunities a 2YC professor has to teach introductory courses in one semester (typically 5–7 per semester). 2YC institutions have more diverse student populations, relative to most geoscience programs at four year institutions (4YC), and as such can better represent a broader perspective of the experiences of minority students. Finally, the 2YC faculty focus predominately on teaching and learning, and less on research and publications, allowing more time to develop effective pedagogy. As we continue to rise to the challenge of improving geoscience recruitment, education, and retention, collaboration between institution types and their faculty are a key component to our success.

LABORATORY LEARNING SETTINGS

The laboratory environment provides excellent opportunities for students to address problems or questions that require them to think and act like scientists. Lab experiences should help our students develop a mix of technical skills (e.g., map reading, mineral identification) and scientific reasoning skills (e.g., interpreting the geologic history of an area from a cross section). Lab activities can vary from relatively basic exercises in introductory courses (e.g., using geologic and topographic maps to answer questions), to more sophisticated tasks in majors' courses such as research, experimentation, fieldwork, or modeling and simulations. Conducting inquiry-based activities in laboratory courses can promote student-centered teaching that emphasizes the process of scientific inquiry and focuses on the student's role in investigating scientific questions and building conceptual understanding. Inquiry-based lab activities in geoscience courses improve student learning in comparison to more traditional lab exercises (Grissom et al., 2015; Moss and Cervato, 2016). Based on the *2014 –2015 Summit survey*, most active learning occurs in course-related labs, with 53% of faculty saying most to all labs in their program are inquiry based (Fig. 5-1).

Inquiry should engage in scientifically-oriented questions, give priority to evidence, formulate explanations from evidence, connect explanations to scientific knowledge, and communicate and justify explanations (NRC, 2000). Students are unlikely to experience fully authentic inquiry (formulate their own research question, design their procedures, analyze and communicate their results, and formulate conclusions) except in selected upper level courses or in specially designed research-experiences. Instead, lab activities should run the from low-level inquiry such as basic confirmation tasks (e.g., students identify a mineral or read a map elevation) to higher level open inquiry tasks that provide students with a question and some background information and expect them to design their experimental procedures and decide how to analyze and communicate their results (Ryker and McConnell, 2017).

The level of inquiry should match the task at hand. A review of activities in four published lab manuals for introductory physical geology courses revealed that most (~88%) of the included activities involved relatively low levels of inquiry. Limiting students to low-inquiry activities may influence their views of the nature of science as a primarily confirmatory, fact-gathering activity, and may negatively influence their perceptions of geoscience. Lower-level inquiry activities are suitable for descriptive tasks while activities with higher levels of inquiry are more appropriate for more complex and abstract concepts. By providing a mix of high- and low-inquiry activities in introductory geology laboratory courses, students develop a better understanding of geoscience and of the nature of science. Students who have positive experiences in introductory level science are more likely to persist and take a second course in the discipline. Although tasking students with completing higher-level inquiry activities may cause frustration, scaffolding the learning for those students so they can engage in these activities can lead to a sense of accomplishment, improved theoretical understanding, and a view of science as a creative process by which we investigate the world around us (Ryker and McConnell, 2017).

Box 5.4: Marine Geology and Geophysics Field Course: Experiential Learning Through Collaborative, Interdisciplinary Team-Based Field Project

Each Maymester, the University of Texas Institute for Geophysics (UTIG) offers a marine geology and geophysical field course designed to provide hands-on instruction for upper-level undergraduate and graduate students in the collection, processing, analysis, and interpretation of marine geological and geophysical data. The class is taught by three researchers with three marine technicians who act as the teaching assistants and oversee safety training and instruction.

Initially, the class provides students with three days of classroom instruction on the theory for the different techniques. The class then travels to the Gulf Coast for a week of at-sea field work and on-shore lab work. Students gather multibeam bathymetry, sidescan sonar, high-resolution seismic reflection and chirp sub-bottom profiling data, and take sediment cores and grab samples. In the lab portion, they analyze their geophysical data and the sedimentology of resulting seabed samples (e.g., core description, grain size analysis, x-radiography, etc.).

Upon returning to Austin, students work in teams to integrate data and techniques into a final project that examines the geologic history and/or sedimentary processes as typified by a small area of the Gulf Coast continental shelf. Students spend one week learning interpretation methods using industry-standard, state-of-the-art software (Focus, Landmark, Caris, Fledermaus). On the last day, students present their final project to the instructors, their fellow classmates, and foundation and industry sponsors.

Courtesy of the Jackson School of Geosciences, University of Texas at Austin

EXPERIENTIAL LEARNING

Experiential learning is an effective way that students gain skills, knowledge, and experience outside of the traditional academic classroom or associated laboratory setting. These experiences take the form of co-curricular activities such as internships, studies abroad, Research Experiences for Undergraduates (REU) and service-learning projects. Field trips and courses, field research, senior or honors theses and independent research projects also provide first-hand experience for students to learn concepts and develop proficiency or mastery of geoscience and science skills. Most experiential learning activities offer opportunities for authentic inquiry. Conducting research develops critical thinking and problem-solving skills while developing a fuller understanding of content knowledge. Field courses and field camps commonly include multiple types of group projects that, in addition to making direct field observations and building team skills, require thinking on geologic and real timescales, spatial thinking in 3D (and 4D), and geologic reasoning and synthesis. Many of these courses also include written or oral presentations.

Another approach to experiential learning is course-based research experiences (NASEM, 2015). Some of these build entire classes around collaborative, interdisciplinary team-based projects, providing experience in project management and teamwork. Such courses are often capstone experiences, like engineering design courses. These courses can incorporate service learning, research projects, or collaboration with local geoscience employers (sometimes associated with internships). Other examples include courses designed around field-based investigative projects (see Box 5.4) or modeled after the AAPG Imperial Barrel competition[17] where student teams do a prospective basin evaluation, analyze a dataset (geology, geophysics, land, production infrastructure, and other relevant materials) and prepare a 25-minute presentation on their results. Some departments have developed a different approach where a large student project (individual or team based) spans multiple courses, and students work on different aspects in each course. In the *2014–2015 survey*, 30% of respondents reported that one or more project-oriented courses were offered in their departments (Fig. 5-1).

17 https://iba.aapg.org/program

ALTERNATIVE LEARNING SETTINGS

Another type of instructional reform focuses on altering the course structure itself by introducing opportunities for learning outside the confines of a typical classroom through the addition of online components. Such changes are facilitated by the prevalence of online learning management systems (e.g., Blackboard, Moodle, Canvas) and the availability of extensive suites of free resources to support geoscience instruction[18], especially in introductory courses. One challenge for instructors seeking to use active learning strategies in their classes is finding the course time to incorporate these strategies without sacrificing content coverage (Crouch & Mazur, 2001). This dilemma has contributed to the development of the "inverted" or "flipped" learning model that places some of the responsibility for learning basic content on students outside of class time, which is often facilitated by the use of quizzes and on-line assessments (e.g., Lage et al., 2000; Bishop and Verleger, 2013; Gross et al., 2015). This ensures students have a stake in interacting with the basic content, and, by moving some content outside of the classroom experience, provides additional class time for students to tackle more demanding activities with the instructor and peers (Strayer, 2012; Gajjar, 2013). Additionally, the practice of "flipping" may allow the instructor to explore topics in greater depth, and present students with opportunities for formative assessments so they can practice applying a new concept and assess their comprehension (Freeman et al., 2011). Implementing such classroom approaches involves substantial initial logistical planning.

18 For example, the Teach the Earth portal at SERC: https://serc.carleton.edu/teachearth

Box 5.5: Examples of "Flipped" Classes

2-year community college:

I have successfully transitioned to a flipped classroom model, which focuses on active learning vs. passive listening. I use only open educational resources (OER) in my introductory courses and have developed exercises that fully utilize classroom time to explore topics through investigation using a plethora of physical samples and models plus technology using large data sets, interactivity, animations, and video instruction. I assure that students leave my introductory courses with the skills necessary to analyze geoscience related data to assist them in their future course of study and in their everyday lives.

Doctoral-granting R1 public university:

Peer mentoring works in active-learning classrooms. "New" faculty co-teach these "flipped" courses with seasoned instructors and then take over as lead instructor after 1–2 semesters. This has been successful with three faculty in our introductory course. With the help of a teaching postdoc, we also "flipped" the associated lab course. Metrics for this introductory course show the transformed format promotes success for underrepresented minorities (URM) and women, and URMs in our major have increased over the time-period of the transformation. We implemented new peer teaching evaluation that includes teaching practices inventory (http://www.cwsei.ubc.ca/resources/TeachingPracticesInventory.htm). Initially, this process was done for an introductory course, however, they are slowly drawing more faculty into transformed courses and these experiences appear to "trickle" up to their major courses. Mineralogy, Geochemistry, and Biogeochemistry are now also taught as "flipped" class. The mentoring approach used within the transformed courses was very successful. Student evaluations for these instructors were immediately high without any of the dips seen with other implementation strategies. A new graduate training course, with specific training in active learning, greatly improved TA support in these types of courses and will be expanded this year to include more strategies for student success (time management, etc.).

USE OF TECHNOLOGY IN TEACHING AND LEARNING

Technology is increasingly used for instruction and to transform the structure of classrooms (Whitmeyer et al., 2016). Creating 3D visualizations and/or animations, either as models of geologic features or from real data (e.g., maps and cross sections), which can be rotated or viewed as 2D sections in any orientation, greatly increases students' ability to visualize and understand complex geologic structures, features, and spatial relationships. Witnessing major geologic events virtually through time-lapsed videos enhances student understanding of geologic processes. The development of virtual field trips and lab experiences makes it possible for many students to have experiences their institutions cannot provide. Thus, internet and information technologies have evolved as instructional tools from "eye candy" into sophisticated systems that facilitate student exploration and classroom interactivity. The major advances taking place in visualization and geospatial tools, generation and use of massive amounts of quantitative information, and computational modeling and simulation provide both predictive capabilities and insight into processes and global-scale events.

In the *2014–2015 survey*, most respondents (94%) indicated that faculty used passive observation of visualizations in the classroom, however, many (>60%) were using technology that actively engaged students in classroom settings. Virtual fieldtrips, student driven investigation with real time feedback, and social networking, games and crowdsourcing had under 31% penetration (Fig. 5-2). With the onset of the COVID-19 pandemic and the canceling of many 2020 field camps and courses, the community has greatly accelerated the development of online field exercises, simulations, and computer "games"[19].

In standard field courses, many institutions already incorporate technology. In the *2014–2015 survey*, up to 60% of faculty respondents said technology was used in conjunction with fieldwork, ranging from mapping on tablets or phones (23%) to using ArcGIS or other similar software (49%), to the instructor (60%) or student (35%) preparing or developing information in advance from remote sensing, Google Earth®, DEMs, etc., prior to conducting fieldwork (Fig. 5-3).

The *2014* and *2016 Summit* participants discussed the evolution, potential, and

19 https://serc.carleton.edu/NAGTWorkshops/online_field/activities.html

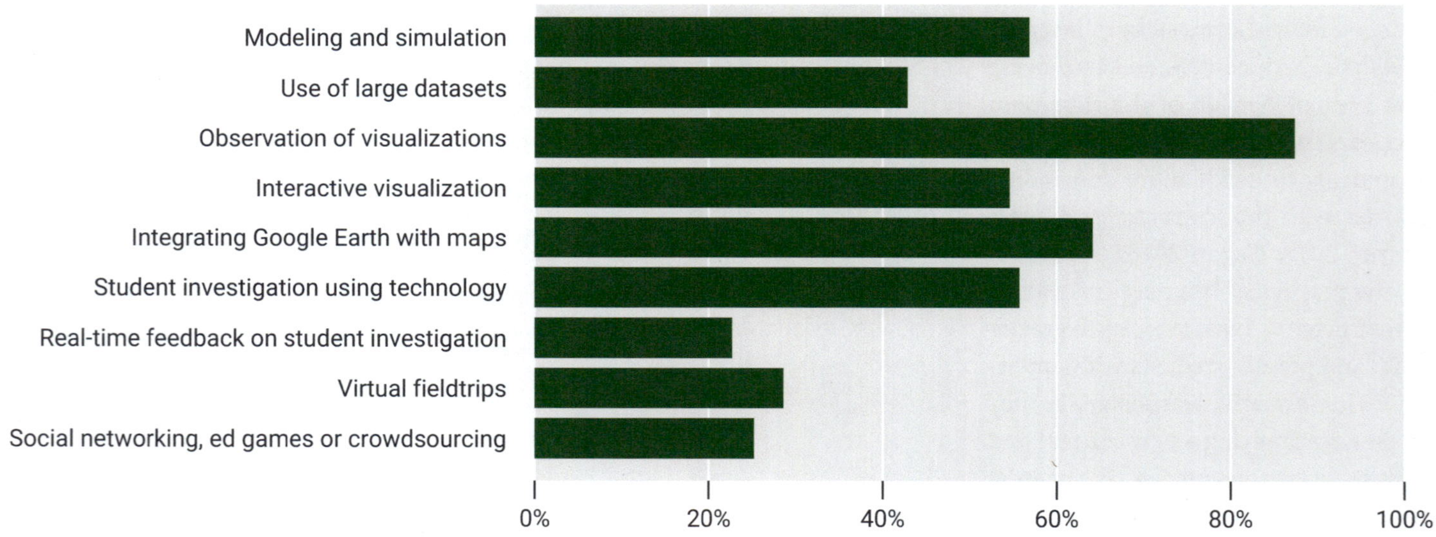

Figure 5-2: Technologies Used in Teaching

Percent of responding departments

Survey Question: From what you know about your department, how do your colleagues (or you) use technology in teaching?

challenges of technology in teaching and learning, and little had changed between the two meetings in terms of usage. With some notable exceptions, the geoscience community has not fully embraced the potential of such approaches, including social networking, educational games, and crowdsourcing, to effectively engage with a generation of students immersed in digital information and online social interactions. Community-wide strategies for developing and integrating appropriate technological advances into future educational efforts (e.g., leveraging the shift to mobile devices, harnessing data for educational research, or developing virtual communities of practice) are needed. One positive outcome of the COVID-19 pandemic has been a significant increase in such development and sharing of such resources between educational institutions with local customization.

Courtesy of the Jackson School of Geosciences, University of Texas at Austin

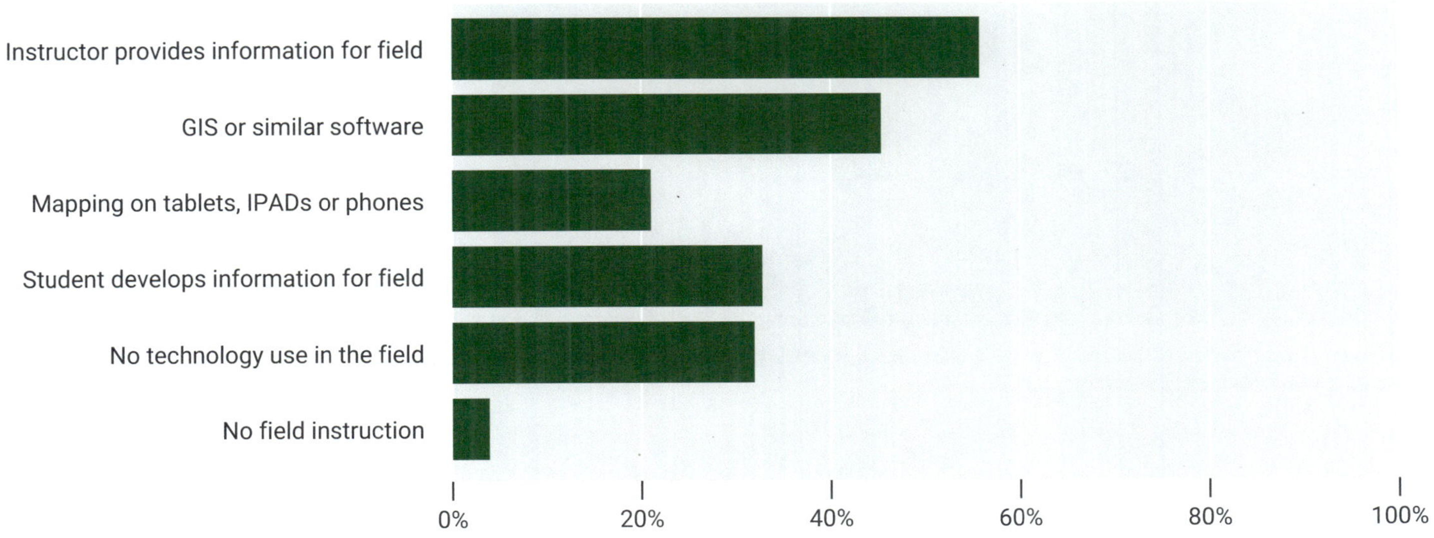

Figure 5-3: Technologies Used in Field Instruction

Percent of responses from departments and faculty

Survey Question: From what you know about your department, how do your colleagues (or you) use technology in teaching in the field?

Maxine Brown, Image courtesy of the Electronic Visualization Laboratory at the University of Illinois at Chicago, for AGI's 2016 Life as a Geoscientist contest

Recommendations:

- Instructors should become conversant with, and adopt, active learning teaching strategies that research has shown to motivate students and enable improved learning.
- Incorporate activities that develop skills and competencies in multiple classes so students can practice, establish mastery, and recognize how these proficiencies are broadly employed
- Conduct inquiry-based activities in laboratory courses to emphasize the process of scientific inquiry and focus on the students' role in investigating scientific questions and building conceptual understanding
- Use current and emerging technology and computational models and simulations using large datasets to increase student understanding of complex geologic structures, features, and spatial relationships and to provide insight into processes and global-scale events
- Request or provide, depending on your role, opportunities for faculty to implement active learning and other pedagogical innovation — such as time, redistribution of workload, non-instructional assignments, and professional development, according to individual institutional policies

6. Practices for Assessment of Student Learning Outcomes

The community-vetted suite of student learning outcomes for discipline-specific and professional competencies can be used as the basis for geoscience program assessment.

Assessment of undergraduate geoscience curricula is the data-driven measurement of a program's effectiveness in supporting student learning across a set of critical areas (Mogk, 2014). A learning assessment strategy is based on the desired student learning outcomes in individual courses and across a curriculum. The first step in developing an effective and tractable learning assessment protocol is identifying the key student learning outcomes. At the class level, instructors should have both course learning goals and related learning objectives for each class session that are matched with appropriate assessments. At the curriculum level, faculty need to agree on the desired overall learning outcomes for the students and develop appropriate assessments. If some desired learning outcomes cannot be met by their program, programs should provide students either co-curricular activities or information on external resources to meet these outcomes.

The geosciences, unlike engineering which must follow the ABET accrediting criteria, do not have externally mandated priorities for the discipline. The consensus findings from the *Summit* efforts provide an *externally-vetted* suite of student learning outcomes spanning the range of discipline-specific concepts, professional skills, and competencies that can be used as the foundation for assessment plans for undergraduate geoscience programs nationally.

Departments that use the backwards design or similar approaches to revise their curriculum will develop a matrix of student learning outcomes that provides a blueprint for assessment. Additionally, when colleagues are aware of prerequisite requirements, they can communicate with peers to help with assessment of individual course effectiveness. Aside from disciplinary, program, or department standards, student learning outcomes for any degree program need to align with institution-level learning outcomes, both to meet institutional needs as well as to enhance the efficiency of the data collection and analysis effort, which can become highly onerous if not approached strategically.

A major challenge to any successful geoscience curricular assessment effort is finding reasonable means for measuring disciplinary, as well as non-disciplinary, professional skills, such as teamwork, critical thinking, written, and oral communication. Several very useful rubrics have already been developed and are readily available for most the competencies. For example, the American Association of Colleges and Universities (AACU) VALUE project[20] has developed an extensive series

OBSERVATIONS FROM THE COMMUNITY:
Leveraged annual Program Review and mapping core competencies onto departmental and institutional learning outcomes; and constructed a competencies matrix to make sure that students acquire core competencies and desired skills through our program. *(Bachelor's-granting private college)*

20 https://www.aacu.org/value-rubrics

Box 6.1: Capstone Field Courses/Camps

What field experiences should aim to accomplish. These field course/camps use most of the skills needed while integrating the concepts learned and allow student to demonstrate their competencies.

Learning Objectives & Assessments for Field Experiences, from NAGT workshop led by Kurtis Burmeister and Laura Rademacher[22]

- Design a field strategy to collect or select data to answer a geologic question
- Collect accurate and sufficient data on field relationships and record these using disciplinary conventions (field notes, map symbols, etc.)
- Synthesize geologic data, and integrate with core concepts and skills, into a cohesive spatial and temporal scientific interpretation
- Interpret Earth systems and past/current/future processes using multiple lines of spatially distributed evidence
- Develop an argument that is consistent with available evidence and uncertainty
- Communicate clearly using written, verbal, and/or visual media (e.g., maps, cross-sections, reports) with discipline-specific terminology appropriate to your audience
- Work effectively, independently, and collaboratively (e.g., commitment, reliability, leadership, open for advice, channels of communication, supportive, inclusive)
- Reflect on personal strengths and challenges (e.g., in study design, safety, time management, independent, and collaborative work)
- Demonstrate behaviors expected of professional geoscientists (e.g., time management, work preparation, collegiality, health and safety, ethics)

22 https://serc.carleton.edu/NAGTWorkshops/online_field/learning_outcomes.html

of assessment rubrics for these kinds of competencies: rubrics are provided for inquiry, critical and creative thinking, oral and written communication, teamwork, problem solving, reading, quantitative and information literacy, integrative and applied learning, and lifelong learning. The *National Association of Colleges and Employers* (NACE) also provides useful student learning rubrics and career assessment resources. Additionally, faculty can assess aggregate student learning outcomes for an individual course using a SALG—Student Assessment of Their Learning Gains survey[21], which is customizable and allows faculty to tailor questions to their courses. The limitation, however, is that it relies on student self-reporting.

With reliable and usable rubrics in hand, the next challenge facing departments is identifying suitable student work products to review for such competencies. The common advice is to look at "capstone" curricular experiences to find assignments and student work products reflecting the full curricular experience. The traditional capstone experience in geology programs has long been geological field camp courses, which require students to use the information, concepts, and skills they learned in classes to solve field-related geoscience problems (Box 6.1). Although field courses are a valuable assessment tool for many concepts, skills, and competencies, other capstone experiences may provide more suitable measures for many professional and computational/quantitative/data analysis skills. Departments need to examine the assignments and student experiences across their degree programs that allow for the review and analysis of all the desired skills and competencies. This review can be augmented by using student e-portfolios as described below.

21 https://www.salgsite.org/

The Association of State Boards of Geology (ASBOG), which administers the Fundamentals of Geology examination as part of the professional licensure process for geologists in 34 states, supports the use of this examination as an exit assessment for graduating students. In Mississippi, all graduating seniors in geology programs take the exam, and it is also required by the University of West Georgia (ASBOG, 2016). Recent state-level initiatives to establish "Geologist in Training" certifications based on successful completion of the ASBOG Fundamentals examination provides new incentives for graduating seniors to take the exam. ASBOG also reports summary results to departments. As regional accreditors (led by the Southeastern Association of Colleges and Schools: SACS) are beginning to require "external measures" in proposed assessment plans, the use of longstanding profession-oriented instruments like the ASBOG Fundamentals exam begin to look more attractive. An obvious challenge with use of the ASBOG exam is cost, as students, or their department, must cover the fee.

One consideration about the ASBOG Fundamentals exam is that it focuses on conceptual geoscience understanding, as opposed to the mastery of important geoscience and other skills (like mapping, field/laboratory data collection and interpretation, data analytics, etc.). Also, the content coverage is comparatively traditional[23], so topics such as Earth system science and climate change are not directly addressed, and there is a focus on the more applied geosciences, such as hydrogeology and engineering geology. Aside from the ASBOG exam, a different range of validated assessment instruments (i.e., test questions and student assignments that have been validated for assessment use) are available from the Geoscience Concept Inventory[24] (Libarkin and Anderson, 2006). Alternatively, Conceptests[25] and Geoscience Literacy Exam questions[26] represent assessments that can be applied in introductory courses. Concept Sketches (Johnson and Reynolds, 2005) provide a means for quick rubric-based assessment of visual conceptual content that is unique to the geosciences.

Longitudinal, post-graduation assessment measures can provide some of the best evidence for the effectiveness of a degree program. Tracking the professional progress of bachelor's recipients through "check-in" surveys, 1–5 years post-graduation, can provide useful perspectives on their academic preparation and any unanticipated skill needs. The challenge in gathering such data is tracking one's graduates, which increases in difficulty with time since their completion. Successful university alumni and development offices have their own means and motivations to track recent graduates and may be able to assist, as can active alumni society chapters. For example, the University of South Florida Geology Alumni Society maintains a professional email network and adds all the current students who participate in their outreach activities. Alternatively, using professional social networking sites (LinkedIn, Facebook groups) offers a way to create and maintain contact with students both pre- and post-graduation. Holding alumni receptions at major geoscience conferences is also a way to keep in touch with graduates.

Another indirect, but useful, resource is engaging with employer partners as

Getty Images/Cultura RF

23 See https://asbog.org/documents/testblueprint2015.pdf.

24 https://geocognitionresearchlaboratory.com/2018/11/20/the-geoscience-concept-inventory/

25 https://serc.carleton.edu/introgeo/interactive/conctest.html

26 https://serc.carleton.edu/integrate/about/gle.html

John M. Wright for AGI's 2017 Life as a Geoscientist contest

members of program advisory boards, or through surveys and interviews. Employers who hire many of a program's bachelor's recipients see the level of preparation and the nature of the professional skills developed in the program and can provide feedback on the strengths of a program, identify gaps that may exist, and note issues graduates may be consistently having in areas such as communication, teamwork, and professional acumen. Access to this type of assessment data involves developing and nurturing relationships with local and regional employers, a task that is often a sidebar in the service and community engagement activities of geoscience departments. Similarly, faculty colleagues at graduate schools can provide insight into how prepared student graduates are for continued education.

E-PORTFOLIOS AS A STUDENT TOOL FOR DOCUMENTING AND SELF-ASSESSING DEVELOPMENT

Students can participate in self-assessment and document their development through their undergraduate program. They benefit from a solid understanding of their own progress with respect to the linkages between courses and the rationale for both geoscience in-course activities and the necessity of supporting courses. Students can 'validate' their understanding of concepts, skills, and competences, starting with class exposure at the basic level, application in a dedicated assignment being the next step, followed by independent application of the skill to a problem showing a fundamental level of mastery. This tiered approach is often used within corporate development processes. Documenting achievements is a core challenge with competency-based approaches. Courses, as a traditional unit, are a relevant benchmark even in a competency-based perspective, especially if the intended learning outcomes are clearly identified to students so they can recognize them, record them, and consider how they integrate into future learning experiences. Likewise, as additional co-curricular activities are integrated into the student experience, these non-traditional approaches also need to find a way into the demonstrated record of the student.

One solution for facilitating student demonstration of achievement across a spectrum of modes is the use of an individual digital portfolio of work, transcripts, and achievements, usually referred to as an ePortfolio (e.g., Cribb, 2018). These ePortfolios allow students craft a professional identity in a digital format similar to a website or blog. Three general types of ePortfolios are widely recognized:

1. A Working portfolio, or "holding tank", where students are building the portfolio in iterative cycles of creating, reflecting, and revising (also called integrated learning or developmental portfolios). The act of creating a portfolio is itself a learning experience for the student and may involve working with a mentor during its development.

2. The Showcase or Display portfolio is where students allow others to view their portfolio to demonstrate their achievements and evidence of mastery, either collectively to potential employers, graduate schools, or for specific projects.

3. The Documentation or Directed Portfolio is generally used for assessment and is more structured around program outcomes. This portfolio facilitates student mapping of work products, achievements, and other accomplishments to the programs intended outcomes.

Departments and programs can use these collectively as a means of assessment. Extensive literature on ePortfolios exists (e.g., Danielson and Abrutyn, 1997; Matthews-DeNatele, 2013; Barrett, 2010[27]), and case studies demonstrate a correlation between the use of ePortfolios and student engagement in learning, retention, success, and catalyzing learning-centered institutional change (Eynon and Gambino, 2017, 2018).

Student ePortfolios can support evaluation of how well a geoscience program's activities are meeting intended outcomes, especially in the complicated dance between the formal educational process controlled by the department, and the needed co-curricular activities for which there is less departmental influence. This situation was well described by Matthews-DeNatele (2013) as ePortfolios "living out the tension between data-driven strategies for verification and accountability, and personalized learning that is greater than the sum of its parts, and thus difficult to measure."

The ePortfolio can become a permanent, dynamic record of the individual's career, from formal education through their lifetime of work and learning. This records the continued development of their individual professional development plan and becomes a reference for completing more traditional records of achievements, such as resumes and curricula vita.

Recommendations:

- Understand and define what constitutes the varied modes of success of your program — post-baccalaureate student outcomes, curricular outcomes, and individual course outcomes
- Have clear learning outcomes and matching assessments within individual courses
- Develop an effective and tractable learning assessment protocol for your program that uses key student learning outcomes and incorporates specific discipline, program/departmental, and university goals
- Use external assessment methods, such as AACU rubrics, ASBOG exams that reflects employer needs, ABET for frameworks, and post-graduation longitudinal surveys to understand skills gaps in the program
- Leverage modern student self-assessment tools, such as e-portfolios, the *National Association of Colleges and Employers* (NACE) student learning rubrics, and career assessment resources, etc., as a collaborative opportunity between students and the program on aggregate outcomes

OBSERVATIONS FROM THE COMMUNITY:

We implemented a requirement that all students in specific geoscience courses complete a 'signature assignment' to be uploaded to an ePortfolio. At the end of each semester, student ePortfolio 'signature assignments' are assessed on several specific criteria based on a university-provided rubric. *(Doctoral-granting public university)*

27 https://www.slideshare.net/eportfolios/eportfolios-in-stem

Dan Scott, for AGI's 2015 Life in the Field contest

7. Preparation for Careers

Departments should help students take a proactive role in their education and co-curricular solutions to develop skills needed for future careers.

Courtesy of the Jackson School of Geosciences, University of Texas at Austin

Preparing undergraduate students for work experiences beyond the degree requirements can be challenging for faculty. The goals of undergraduates are diverse with only about a third continuing to graduate studies (Wilson, 2019), the development pathway with which faculty are most familiar. Additionally, few employers provide active career management for their employees, and commonly the best available advice to students is generalized educational pathways for a discipline rather than a specific occupation (Asher et al., 2018) or occupational handbooks such as those produced by the U.S. Bureau of Labor Statistics. Supporting the spectrum of potential pathways, even within the geosciences, is a challenge, let alone for the 20% of geoscience graduates who decided to pursue careers outside of the discipline (Wilson, 2018).

Discussions catalyzed by the *Summits* focused on encouraging geoscience students to take a proactive role in their education, including co-curricular solutions to foster key skills. *Summit* participants noted that their undergraduates were surprisingly unaware of the career possibilities with a geoscience bachelor's degree, and this was an area where actively partnering with student advising on their campuses and with their local, regional, and other employers would be beneficial.

One method to empower students in their own career development management is to have them develop an Individual Development Plan (IDP) with help from a mentor or advisor. These IDPs are generally used for postdoctoral fellows and graduate students[28] but are becoming more common for undergraduates[29]. Such plans provide a customized roadmap for professional training and goals. The focus is on identifying career goals, the skills and competencies required to achieve that goal, the pathway of activities that develop them, and establishing a record of activities to assess progress.

One challenge for geoscience undergraduates attempting to use this planning approach is that the geosciences do not, as of 2020, have a universal detailed skills/competency matrix such as those found in the Department of Labor Career OneStop

28 For example, https://myidp.sciencecareers.org/ or https://www.feinberg.northwestern.edu/sites/ctmh/docs/idp-worksheet.pdf

29 For example, https://medium.com/stem-and-culture-chronicle/building-your-individual-development-plan-idp-a-guide-for-undergraduate-students-f14feca9111c (Bosch, 2017; Sacnas — Stem and Culture Chronicle) or https://undergrad.ucf.edu/whatsnext/faculty-staff/resources/individual-development-plans-idps-for-undergraduate-students/

OBSERVATIONS FROM THE COMMUNITY:
I have written a Geology Roadmap that is given to all majors. This document outlines discipline specific skills and student learning outcomes. *(Bachelor's-granting private college)*

site for many other professions, including geospatial sciences.

In general, students start the process by assessing their own skills and exploring career pathways that are of interest. Once they have identified their career interests and current skills, they determine which skills they need to meet their career goals and establish a development plan. Additionally, students identify where and how they can obtain these skills and what professional development outside of their classes is needed. Integrating their IDP with an e-portfolio is one way to track and demonstrate their progress.

Developing their IDP with help from a mentor or faculty advisor throughout their education is important. Goals and interests change as students learn, and students cannot always develop the skills required for a specific career. These plans should facilitate agile responses to progress or changes in goals or the target environment, such as fundamental structural change to an industry. Although many students do not come into the geosciences with a specific career goal, some do but are unaware of the required skills and competencies. In advising these students, it is important to help them identify what they will need to learn and be able to do to follow that career path. The student will need to analyze whether they wish to pursue that career based on this information and their own skills, values, and interests, or whether to investigate and pursue other career paths.

Geoscience programs that have detailed the learning outcomes of their courses can provide the critical framework for the student to begin mapping their plan. With help from faculty, a student with access to the menu of learning outcomes available in a course of study can begin to map the portfolio of skills and competencies needed to support the trajectory towards their next career step. If it includes graduate school, the student needs to take into consideration the knowledge and skills they will need to be able to get into graduate school in their chosen field (e.g., field experience, high-level math, or computational methods).

In addition to technical skills and competencies, the mapping process should include professional skills such as writing and oral presentation, budgeting, project management, and, if appropriate, regulatory certification. Faculty advisors can be key in helping students understand when they are developing these skills, which may not be mapped to a specific course outcome and may be experiential or co-curricular in nature. For instance, designing and executing a senior research thesis study develops project management skills which should be identified to the student (Wulff and May, 2013). Likewise, faculty can help students understand that achieving a skill is not a binary experience, but rather one in which they progress from exposure, practical application, competency, and finally mastery. At the same time, students should understand they do not *always* need to be the master of *all* skills.

Some programs have started to implement student e-portfolios in which students compile examples of their curricular and co-curricular work products to document their educational experiences and developing skillsets for future employers and/or graduate programs (Cribb, 2018; see Section 6). These electronic portfolios can be reviewed by faculty to provide students guidance on how to best present their strengths. Additionally, many universities are developing stackable credentials, such as badges, or certificates for completion of a program of study or series of learning experiences that are awarded for mastery of a specific competency, which can be incorporated into e-portfolios. Many of these are designed to be relevant to specific careers, for example an energy management certificate or data analytics certificate.

A mechanism to improve students awareness of the skills and competencies needed for various careers is to have students interview individuals in occupations close to their goals about how those individuals rose to that position. This is a methodology used in some European energy companies to help new employees map their desired career trajectory into management or senior technical roles (Rosaz, 2013). In addition to the typical curricular benchmarks that will be reaffirmed, such processes help students identify not only the additional needed skills, but also the diversity of methods of acquiring those skills.

Students who participate in career-related activities, such as co-ops, internships, or extended problem-based learning experiences, gain first-hand experience with potential careers. Departments that develop collaborative relationships with employers to foster these types of opportunities for students also strengthen their program's connection to industry and enhance their relevance on campus and in their community. The depth of involvement of departments with employers is highly variable, but whether sustaining or developing these contacts, overcoming the challenges of sustaining these external relationships do help students establish a network of potential employers.

To support the diverse career interests of geoscience students, geoscience professional societies are also providing a number of services: the American Geoscience Institute (AGI) Career Compass[30] effort provides "roadmaps" for students and advisors to the critical skills and experiences that undergraduates interested in pursuing careers in a wide-range of professional directions should seek. The Geological Society of America (GSA) GeoCareers programs include the Schlemon and Mann Mentoring programs at GSA professional meetings where students can interact directly with professionals across a range of industries. The American Geophysical Union (AGU) has partnered with other societies to establish Mentoring365, a virtual mentoring program providing mentors from outside of academia. The NSF-funded SAGE 2YC project[31], hosted by the Science Education Resource Center (SERC), has developed an extensive suite of geo-career focused web resources aimed at freshmen and sophomores.

Students need to know what careers exist, where to search, what their options are, and how to leverage their competencies, skills, and knowledge to get an interview and be hired. Departments and programs can help guide students to job-seeking resources, whether within the department, somewhere within the institution, or through external groups like societies. Job seeking support can include providing advice on resumes, applications, and interview skills, which is often available on campus or even some employers who hire students from a program or local alumni are willing to provide in-house workshops.

Departments need to be intentional in nurturing and leveraging their connections with geoscience employers to support geoscience undergraduates through their degrees and post-graduation (Box 7.1). These employers can provide feedback on department programs through advisory boards and offer valuable student co-curricular professional experiences through internships, field experiences, and various kinds of professional development workshops and/or short courses. Geoscience professionals who participated in the *2015 Employers Workshop* also noted the potential of some to support undergraduate research activities by providing datasets for analysis, or other kinds of hands-on field and/or lab experiences. Constructively engaging with geoscience employers and alumni involves persistent and creative engagement of one or more faculty members and is something that often goes unrewarded academically. However, as noted by Boyer (1990), constructive and successful outreach efforts are themselves a form of scholarly activity. To transform geoscience undergraduate education, these kinds of activities need to be recognized as mission-central and scholarly and be supported and rewarded by heads/chairs and administrators.

30 https://www.americangeosciences.org/workforce/compass

31 https://serc.carleton.edu/sage2yc/careers

Box 7.1: Career-focused Geoscience Courses at the University of South Florida

The product of a vital and evolving Department/Employer partnership (see Ryan and Schackne, 2016; Ryan et al., 2017)

The University of South Florida (USF) Geology Alumni Society, founded in 1997, serves as the liaising entity between the USF Department of Geology (and more recently the USF School of Geosciences) and the professional geoscience community in Florida, and is an active partner in delivering key pieces of both USF's graduate and undergraduate geoscience degree curricula. At the undergraduate level, regional employers are central players (as presenters and as a co-instructor) in a new 2000-level course, "Preparing for a Career in the Geosciences", which was added to the USF Bachelor's degree offerings based on the recognition that many students were graduating with no clear idea of the professional opportunities in the field. In the course taught by a practicing geoscience professional and past Alumni Society leader, students interact with panels of 2–4 professionals from a range of geoscience employment sectors, many of whom are USF Geology alumni. The professionals introduce their work and background, and students interact with them through extended question-answer sessions. Student assignments involve writing reflective commentary on each of the panel discussions, and an informal professional development plan for their time at USF, based on their interests and what they have learned in the class.

Also, the Alumni Society offers a range of co-curricular professional development and networking opportunities for USF geoscience undergraduates, from social events, to workshops on resume writing and career opportunities, panel discussion events on professional licensing, and related topics, and live field demonstrations during periodic "USF GeoExpo" events, where local firms drill wells, conduct geophysical surveys, and do water sampling at an on-campus "geo-park" maintained by the School of Geosciences. Florida geoscience employers, through the USF Geology Alumni Society, also are intimately involved in the USF Professional Science Master's degree program in Geology.

The USF Alumni Society's involvement in the student's level of education is motivated by seeing them as future employees. Geology Department/School of Geoscience faculty work to provide access by helping reduce administrative barriers to course involvement, and actively reach out for input and feedback on our programs. Much of this work is done by a few department champions who were given assigned time by chairs to network with alumni and employers, forge these connections, and occasionally support events if funding is provided. Also important in maintaining this employer partnership is clear communication about how things work in the context of a large state-funded university (e.g., the time it takes to get a program or course approved, or an on-campus demonstration event cleared to go forward); recognizing the limitations of what employers can and can't do (ex: holding the above described classes only in the evenings so no one has to miss work to participate); and expressing sincere gratitude to all participating employers, both formally (through letters from Chairs, Deans, and other University officials) and informally.

Recommendations:

- Collaborate with students to customize a roadmap for professional training and goals based on skills and career interests (e.g., IDP, e-portfolios, external certifications, and professional development)
- Help students find the resources they need to explore career options, find employment, and get training on the application and interview process, including support from institutional career centers, employers and/or alumni, professional societies, and other professional development resources

8. Recruiting, Retaining, and Promoting Success of Undergraduate Students with Emphasis on Students Underrepresented in the Geosciences

Growing demand for geoscientists requires departments and programs to recruit, retain, and promote the success of undergraduate geoscience majors across a broad spectrum of society.

Courtesy of the Jackson School of Geosciences, University of Texas at Austin

CHALLENGES AND OPPORTUNITIES

The demand for geoscientists in a range of employment opportunities continues to expand and outpace the number of students preparing for geoscience careers. The American Geoscience Institute (AGI) using U.S. Bureau of Labor Statistics data estimates that by 2028, there will be a shortage of 35,000 geoscientists (FTEs, full time equivalents), and geoscience employment will increase over the next decade by 4–8%, depending on the specific occupation. Traditional undergraduate geoscience enrollments have declined recently with six major universities reporting between 28% and 71% (54% average) declines, and many smaller colleges report 'plummeting' or steeply declining enrollments (data from Summit action plan progress reports). Recruitment of students and retention to degree completion is critically important to our profession.

The demographics of the general workforce has changed, becoming more diverse and global, yet the geosciences enroll one of the lowest percentages of underrepresented minority students of all STEM fields (6%; Gonzales, 2010). The need for increasing diversity has been recognized for decades (e.g., initiation of NSF Opportunities for Enhancing Diversity in the Geosciences (OEDG) program in early 2000s (Karsten, 2019); commencement in 1974 of AGI's Minority Participation Program Scholarships) and is affirmed by position statements of geoscience professional societies[32]. The geosciences are still not effectively engaging the entire student population and thus are not competing for the best minds (Bernard and Cooperdock, 2018; Hofstra et al., 2020). Not only are we missing excellent future geoscience professionals, but also the diverse life experiences and perspectives that help in identifying and solving the geoscience-related problems facing society.

Most middle and high school students are unaware of what most scientists actually do, partly because they are rarely exposed to the active work of scientists. Most K–12 students have limited exposure to the geosciences, which impacts their building knowledge of our planet, its issues, and the potential of a career in the geosciences. Evolving science education standards (e.g., Framework for K–12

32 GSA https://www.geosociety.org/gsa/positions/position15.aspx; AGU https://www.agu.org/Learn-About-AGU/About-AGU/Diversity-and-Inclusion

OBSERVATIONS FROM THE COMMUNITY:
2YC: I have developed a number of relationships with local community organizations that have created a number of undergraduate research opportunities for students in a service learning capacity that is helping build community in the classroom, students develop real-world skills, interact, and build relationships with their local community. *(2-year community college)*

Science Education; NGSS) are improving these factors nationwide. Pathways for attracting high school graduates to careers in the geosciences include tapping into an interest in the world around them, working towards constructive societal impact, leveraging interest in high tech fields, and seeking well-compensated careers. Geoscience programs must engage these students as they enter post-secondary education, and work within the community of campus programs to foster interest in the discipline, both for the health of the discipline and the potential to engage the best and brightest.

The geosciences, like most STEM fields, struggle with recruiting and retaining diverse populations through graduation and actively promoting their success, not just completion of their degree. Developing a diverse workforce begins with emphasizing the role geoscientists play in societal and environmental issues and their role across society, providing tangible context on the world of geoscience work. Climate change, sustainability and availability of natural resources, and the increasing impact of natural hazards due to population growth are all beginning to affect society on larger scales. Though no community is spared, the increasing impact of these issues often disproportionately affects underrepresented minorities. If the geoscience community pulls from the greatest breadth of society, these diverse perspectives will lead to unique insights and solutions and effective engagement with all affected communities.

Limited opportunities for being outdoors, in outdoors spaces, or outdoors-related programs impedes the development of curiosity and discovery of the natural world and is a growing challenge to engaging potential geoscience students, and it is particularly acute for many underrepresented and urban populations. Communities, school districts, and institutions of higher education need to create opportunities for outdoor experiences so students can see the benefits and the impact of the environment on their life and local community.

Geoscience departments need to be on the forefront of marketing the geosciences as a fruitful and impactful career that benefits society and local communities. Departments and programs need to directly engage with prospective students and parents, as well as the campus-wide community, highlighting the local, societal impacts of the geosciences and its career prospects by developing community service activities that increase its exposure and build robust STEM support and mentoring systems for students. Faculty involved in such efforts should be rewarded for their contributions to this critically important challenge. We must ensure that we demonstrate the geosciences as a viable, honest, respectable, and intellectual occupation in our efforts to engage with underrepresented minority families.

Geoscientists work with people from all parts of the globe, from urban populations in developed countries to isolated and impoverished villages in the developing world. Students must be prepared to work effectively and respectfully in a culturally diverse environment. Experiences working with diverse populations while in college is a crucial step in that learning journey, so geoscience programs need to reflect and engage with a demographic representative of the nation and humanity. We need to ensure that different cultural perspectives, learning opportunities, and experiences are integral to the geoscience educational process.

Building a representative geoscience community has been a slow process, not from a lack of desire or effort, but because of the complexity of engaging at multiple societal, racial, cultural, and educational levels. The geoscience community needs to take a holistic approach to increasing diversity and a long view. It will take more than one generation to see full representation in the discipline. Consider that an 8th grader in 2020 will be mid-career in 2050, so structural changes we make today in early education will only become inherent in the community in the second half of the century. We must make concerted efforts to set these internal processes in motion now, and look for near-term opportunities, such as encouraging bright undergraduates on our campuses to become geoscientists, to mitigate the immediate issues.

SUMMIT OUTCOMES

The *2014 and 2016 Summit* participants provided valuable insight into best practices for recruiting and retention of students, with an emphasis on students underrepresented in the geosciences, and for successful transitions from two year (2YC) to four-year colleges/universities (4YC). The *2014–2015 survey* indicated that, of the respondents, 40% of departments and 34% of companies and other organizations have, or plan on, systematic efforts to encourage broadening participation and retention of a more diverse student population. Efforts ranged from using role models to collaborating with minority serving institutions (Fig. 8-1). Also, 57% of the departments and 31% of companies and other organizations track the participation and retention of minorities in their population (see Appendix A).

About 22% of the Heads and Chairs who participated at the *2016 Summit* and subsequent workshops submitted progress reports that described successful recruiting and retention strategies. To recruit more students, some departments instituted new courses or changed introductory classes to active learning, worked with college admissions offices on recruitment, held open houses, and used updated marketing materials to advertise their programs to reach wider audiences, including presentations during introductory classes about the geology major and employment opportunities. One department deliberately started co-emphasizing laboratory and computer (e.g., GIS) work alongside fieldwork.

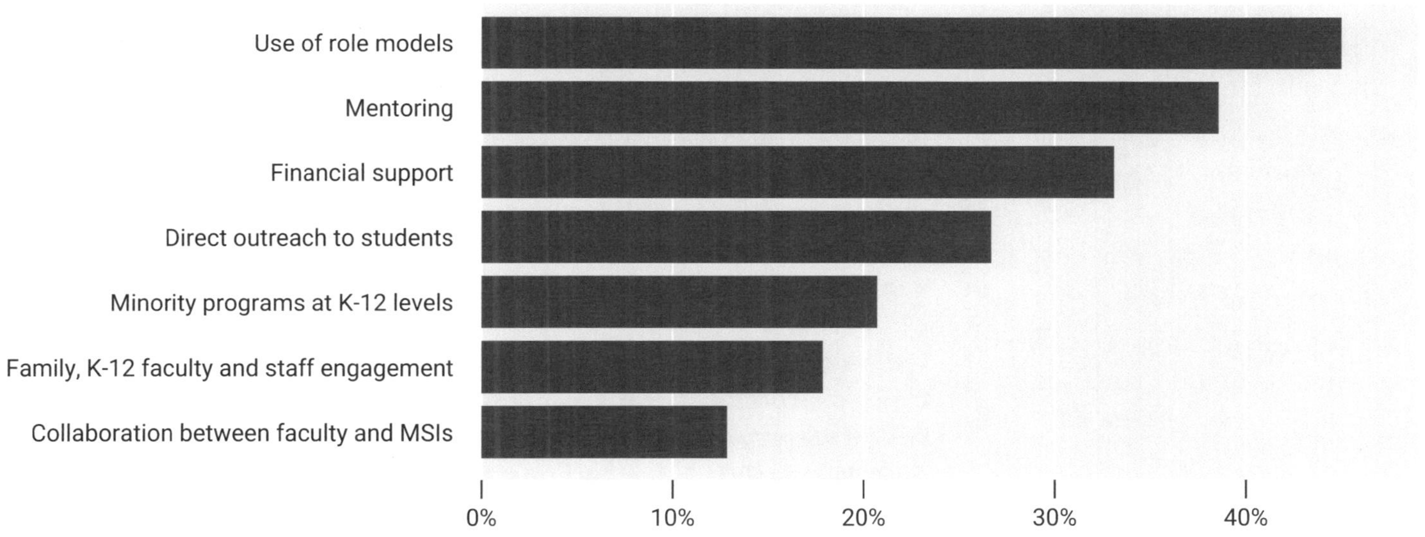

Figure 8-1: Engagement Methods to Broaden Participation

Departments utilizing specific methods of engagement

Survey Question: If you sponsor, partner with, or have a program to encourage broadening of participation and retention of a more diverse student population, which of the following does that program offer?

OBSERVATIONS FROM THE COMMUNITY:
"To increase retention of first-gen college students, we designed a 1-credit course for all students that includes problem-based field and lab activities, discussions, and visits from alumni and industry-professionals. Students will leave with a clear path to graduation, an outline for a portfolio and resume, and knowledge of various career opportunities." *(Bachelor's-granting private university).*

OBSERVATIONS FROM THE COMMUNITY:
We are revamping our freshmen-level classes. Recognizing that classic or traditional freshmen geology classes are becoming less attractive, we were motivated to be proactive in what we teach and had the willingness to do our best to offer modern geology courses in our curriculum. *(Doctoral-granting R1 public university)*

To increase retention of first-generation college and other underrepresented students, several departments increased mentoring of students and became more inclusive of disparate levels of science background by being flexible about the order a student might take science and math courses. One department showed students how courses connect by developing a curriculum roadmap that outlines expected skills and student learning outcomes, and developed e-portfolio programs for student self-assessment. Another department designed a 1-credit course that includes problem-based field and lab activities, discussions, and visits from alumni and industry-professionals.

RECRUITMENT

Recruitment of undergraduate students to the geosciences commonly occurs when undergraduates discover the subject by taking an elective geoscience course. Having the best and most engaging instructors teach these courses is essential; these instructors should use best practices for active learning and concentrate on broad concepts, processes, and human interactions with the Earth system (see Sections 5 and 9).

Integration of practices recommended by the Next Generation Science Standards (NGSS) will resonate with many incoming students from their high school experience (see Section 9). As noted in the Best Practices for Instruction section (Section 5), by having a mix of high- and low-inquiry activities in introductory or non-major laboratory courses, students have more positive experiences and develop a better understanding of the geosciences. These students are more likely to persist and take a second course in the discipline. Math, physics, chemistry, and biology should be integrated into the lower division courses so that students understand why geoscience majors need a spectrum of STEM courses to be successful.

Many students who are initially attracted to the discipline through elective geoscience courses do not realize the technical nature of the geosciences. Departments need to provide prospective majors frank advice on degree requirements, including costs, time commitments and resources needed for the program (e.g., field camp related). Departments can also explicitly target beginning students planning to major in other STEM fields by emphasizing the rigorous aspects of the geosciences and showing that students can use their quantitative and technical skills to make a difference to society while utilizing a broad spectrum of scientific knowledge and skills.

Ultimately, individual departments and programs need to be actively engaged in their own recruitment of students, both on and off campus. Take advantage of any opportunity to promote the geosciences and your program. Collaborating with recruiters in the central admissions office is also important, as they may know little about the geosciences. Provide the recruiters with information on what the geosciences and your program have to offer students. Other possibilities include leveraging other institutional recruiting organizations, such as those associated with athletics, and partnering with other departments, 2YCs and universities.

Websites serve as the front door for the department and program. The site needs to clearly communicate that geoscience is an undergraduate major with a lot of career potential (using AGI data) that impacts their community while solving socially relevant problems, including environmental issues. Websites should be inclusive in their language and imagery. The website is also a key entry point for internal recruitment of undecided undergraduate students, or those enrolled in degree programs that no longer interest them. Include clear roadmaps to degrees, including those for students transferring from other majors and 2YCs. If appropriate, discuss bridge programs starting from K–12 to 2YC to 4YC. Be aware that highlighting students outside in exotic locations working on non-socially relevant problems can be counterproductive.

To engage K–12 students early, representatives of the department or program need to visit schools regularly, especially for career days. Faculty and/or students should talk about different geoscience topics in K–12 classes, judge science fairs, and/or hold social recruiting events. Presentations and media should show diverse geoscientists as successful professionals. It is best to show geoscientists in the lab, on a computer, or in a formal office setting, in addition to the field. Provide career information, such as the American Geoscience Institute workforce brochures and The Earth is Calling (Be a Geo Video[33]), which offer brief introductions to geoscience careers. Visits to schools also provide the opportunity to develop good relationships with high school counselors and provide them with career information. In addition to working with K–12 schools, building stronger relationships with local informal science centers, museums, and civic or professional groups is another way to increase exposure of the geosciences and your programs.

Many departments and programs also work with high school teachers, providing support through educational resources, professional development, fieldtrips, or

33 https://www.youtube.com/watch?v=naMxvhPdi5g

OBSERVATIONS FROM THE COMMUNITY:

We have modified our degree plans, developing two tracks that lead to a degree in Geology. The first track is labeled our "Career Track" and will prepare students for either graduate school or employment in the geology field. The second track is labeled "General Track" and provides the student with a good background in the geosciences, but is probably lacking in some of the cognate courses (specifically Calculus, and less rigorous Chemistry and Physics) that a graduate school or geoscience employer may require. This second track was developed for students who enjoy geology but may end up not working in the geoscience areas. Both tracks have increased flexibility for electives and will allow a smoother path to graduation or allow more specialization for those majors who desire it. It is only one year old, so we are still evaluating how successful this change has been. We think these tracks make the Geology major more attractive to students and our enrollment is slowly, steadily increasing. Many of our new majors are transferring from other majors (Engineering and Physics are key contributors). *(Doctoral-granting public university)*

Courtesy of the Jackson School of Geosciences, University of Texas at Austin

other opportunities. Getting more geoscience content and examples into middle and high school courses builds student exposure. Some departments even offer online geoscience instruction to institutions, particularly to those serving underrepresented communities. Dual credit high school Earth science courses, where students receive both high school and college credits, are another mechanism for developing prospective geoscience majors. Another approach is to mentor students entering international science and engineering fairs (Intel, Siemens, Regeneron, etc.) or Science Olympiads.

As parents often are the most influential factor in their children's decisions related to choosing a higher education pathway (Noel-Levitz, 2009), connecting with them is critical. Parents should be invited to student recruiting events and provided with career, salary, and employment information. Recruiters should be candid with them regarding financial considerations for the degree, including unique expenses like field camp, financial aid, and opportunities for geoscience-related scholarships.

OBSERVATIONS FROM THE COMMUNITY:

We examined the 'face' that geology shows to prospective students. Our revised recruiting and advertising materials have been adapted to reach a wider audience and we have worked with our admissions office to broaden the ways in which they portray geology when talking with students. Specifically, we are deliberately co-emphasizing laboratory and computer (e.g., GIS) work alongside fieldwork. We are inclusive of students coming into our program with disparate levels of science readiness and are flexible about the order a student might take cognate courses, for example. Our numbers are small, so it is difficult to tell whether diverse students find this approach more engaging, but we are paying attention to the implicit messages we send to our prospective students and majors. *(Bachelor's-granting private college)*

We have developed presentations for our introductory classes, informing students about the Geology major and employment opportunities, and this seems to be working. *(Doctoral-granting public university)*

Collaborating with the community, including a local museum and city government, increases outreach and education beyond the community college's walls. This, of course, improves recruitment and retention of students. *(2-year community college)*

Increasing Diversity

The geosciences face major challenges in attracting undergraduate students, particularly those underrepresented in geosciences (Wilson, 2018). Although many STEM fields have similar issues, research has documented specific challenges facing the geosciences (Karsten, 2019), as well as solutions (Wolfe and Riggs, 2017; Gates et al., 2019). The public does not have a clear perception of what geoscientists do, our impact on society, or what geoscience occupations exist. Saddled with stereotypes of boom-and-bust petroleum industry cycles and that most Earth science courses taken in high-school and the introductory college-level are considered an easier science credit, our "storefront" provides little incentive for talented individuals to look more deeply at the geosciences. The lack of authentic geoscience role models who interface with the public over real issues affecting communities leads to misconceptions of what the geosciences are and the breadth of geoscience occupations that are available.

In marketing the geosciences, it is important to emphasize the ability for students to make a difference by solving problems of societal importance. Stressing ties to the local community and the societal aspects of problems is especially important because many underrepresented minorities and first-generation college students view returning to help their community as a major priority (Banks-Santilli 2015). Information on salaries and employment rates are also critical for demonstrating that geoscience occupations are well-compensated and have robust employment opportunities. Another attraction is that the geosciences are engaged in innovation and advanced technologies. When possible, programs should publicly showcase the innovation and advanced technology applications that your program is using while framing the geosciences as a professional occupational discipline.

Two popular geoscience marketing and recruiting approaches, working in the field and global travel, are very attractive to some cultures but not others and need to be presented carefully (see Sherman-Morris and McNeal, 2016). Students who love the outdoors are attracted by the opportunity to work in the field. Similarly, the opportunity to travel and work globally is very attractive to some students. However, many minority communities equate fieldwork with occupations involving manual labor, something they do not want their children doing. The message needs to be clear that geoscientists work as professionals and that most work is accomplished on computers, in offices, and/or in laboratories and that working in the field for a living is a choice, not a requirement.

For community-centric cultures, travel can be viewed negatively by parents and prospective students. It is also important to show that there are opportunities, depending on their career choices, to stay in or near their communities and give back through their work.

The geosciences should adopt the messaging strategies of engineering (National Academy of Engineering; NAE, 2013; chapter 1) which for decades have coordinated with major corporations, the Public Broadcasting System (PBS), the National Academy of Sciences and Engineering (National Academy of Engineering, 2008), and professional marketing firms to present attractive, diverse images of women and minorities in engineering. Lastly, the geoscience community should advertise and demonstrate with tangible local examples that the geosciences can be a gateway undergraduate major that leads to a spectrum of careers by showcasing alumni.

Diversity issues are complex and recruitment or retention strategies will be unique to each institution. Different institutions and regions offer different challenges, potential impacts, and targets (e.g., high schools, two-year colleges, minority serving institutions, other undergraduate majors, etc.). Underrepresented student diversity classifications may include race, ethnicity, first generation, socioeconomic class, gender, age, disability, and veterans and therefore require different approaches.

Understanding the backgrounds of potential students is crucial to using the appropriate tactics and language their specific communities and cultures. Nonetheless, there are some fundamental outreach practices and program elements for improving diversity that can be emulated (discussed at *2014 and 2016 Summits*). Recruitment often starts early during middle and high school with relationship building involving teachers, school counselors, families, and the community. In many cases this contact comes in the form of a geoscience or STEM program for underrepresented minority students at pre-high school and high school levels, either within the community or at the university or college. Whoever is involved in such programs or other recruiting efforts needs to value, and be aware of, cultural differences, local issues, and the roles of different individuals. Personal touch matters. Respect and trust come through building long-term relationships, so ensuring that the same person serves as the "recruiter" over multiple years is important.

Courtesy of the Jackson School of Geosciences, University of Texas at Austin

Involving pre-college students in research programs has shown to increase self-efficacy and a continued interest in geoscience careers (Baber et al., 2010). Summit participants shared information on several successful programs that attracted and retained minority students (Box 8.1). The most successful recruiting programs provided financial support, reached out to students in their communities, involved members of the community (families, high school teachers, guidance counselors), included mentoring, and incorporated role models.

Some successful recruitment programs also provide mentors for prospective students who stay in touch through social media, email, etc. Many underrepresented students are first generation, and no one in their family has applied for colleges or financial aid. A few programs provide workshops to help with admission and financial aid application forms (including Free Application for Federal Student Aid — *FAFSA*®) or offer SAT or ACT preparation workshops. Many offer financial support for students accepted into their programs.

Because most undergraduate institutions' student body is more diverse than most geoscience departments and programs themselves, active recruitment from the entire campus population is another strategy to increase diversity. This recruitment might be achieved through increasing interdisciplinary courses and activities, offering engaging non-major courses, and working across academic departments at universities to give geoscience departments more visibility to a greater range of students.

When recruiting, departments should involve people of similar cultural backgrounds, particularly people closer to the student's age, to more easily establish a rapport. If you have alumni or current students from an underrepresented community, invest in having them return to talk about their experiences. When possible, have minority geoscientists visit the schools and participate in recruiting events as role models.

Diversifying the faculty also demonstrates your commitment to diversity and provides role models and mentors to connect underrepresented students to geoscience careers more effectively (Archer et al., 2019). At the same time, do not overtax faculty from underrepresented populations with recruitment obligations at the expense of meeting their own career goals.

Courtesy of the Jackson School of Geosciences, University of Texas at Austin

Box 8.1: GeoFORCE Texas — Successful High School Diversity Field Program

GeoFORCE Texas[34] is a highly successful K–12 outreach program designed to increase the number and diversity of students pursuing STEM degrees and careers, especially geoscience, at the University of Texas at Austin Jackson School of Geosciences. Each summer, GeoFORCE Texas takes over 300 high school students on geological field trips to the Gulf Coast, Mt. St. Helens/Pacific Northwest, Grand Canyon, and central Texas. These field academies engage diverse students from challenged high schools in southwest Texas and inner city Houston and provide life-changing learning experiences at some of the most spectacular geologic sites in the country to broaden students' understanding of the Earth, geosciences, and engineering. Although it varies each year, the demographics are ~85% minorities (e.g., 2019: 59% hispanic, 17% black, 8% asian) and 60–64% female. GeoFORCE Texas also serves first-generation students and those from low-income families.

Each academy is about 1 week in length and involves active learning in an outdoor environment. Students are recruited in 8th grade and go on one field experience the summer before each of their high school years. They must maintain a B average during the school year and pass quizzes and exams during the week-long trip. Over 1,300 students have completed the program, and 100% graduated from high school. The academies are sponsored by companies and foundations, predominantly the petroleum industry, and are free for the students. Each academy has an instructor, a GeoFORCE coordinator, six counselors, a corporate/professional geoscientist mentor, and an educational coach. Many alumni of GeoFORCE come back and work as counselors in the summer.

GeoFORCE works closely with communities, high school counselors, and teachers. GeoFORCE staff are active with the students during the year, staying in touch and helping them prepare for college applications regardless of what school or major. Staff hold transition-to-college workshops for students and parents on the basics of college, the application and admission process, financial aid and scholarships, and SAT preparation. They also provide letters of recommendation, notify students of potential scholarships, and connect the high school seniors with undergraduates for advice. High school seniors present posters at the Jackson School Student Research Symposium. Incoming college STEM majors also participate in a Math and Science Institute to prepare for their college-level courses.

As of 2019, GeoFORCE had 582 graduates enrolled in college (432 in 4YCs, 77 in 2YC, and 73 in graduate school). ~86% of graduates go onto college with ~90% persisting through their second year. ~44% of undergraduate majors are pursuing a STEM major and 8% more in health and clinical sciences. The geosciences have 8% of the undergraduate and 18% of the graduate students. As of 2019, 492 students have bachelor's degrees, 51 have master's degrees, and 7 professional degrees. 13% of all bachelor's degrees are in the geosciences, 67% of which were earned by underrepresented minority students. Comparatively, 12% of bachelor's degrees in geoscience were awarded to URM in 2016 nationally (NCSES, 2016). Of 51 master's degrees earned by GeoFORCE alumni, 22% have been in geoscience (11).

The staff maintain contact with all the college students, helping them get involved in mentoring or other helpful programs at the university they attend and directing them to research opportunities and scholarships. The Jackson School has endowed scholarships and fellowships for GeoFORCE graduates who are accepted into the undergraduate or graduate program, and geoscience majors who attend universities other than UT Austin frequently participate in undergraduate research with JSG researchers as visiting scholars.

34 https://www.jsg.utexas.edu/geoforce/

RETENTION AND SUCCESSFUL PROGRESS TO GRADUATION

Successful progress to graduation goes beyond retention, particularly for students underrepresented in the geosciences. For example, students may complete the degree by meeting minimum graduation requirements but find themselves lacking the grade point averages and extra-curricular activities sought by graduate schools and employers. For this reason, a multifaceted, institutional approach that nurtures the academic, social, and professional development of all students is central to success (Wolfe and Riggs, 2017). As students declare a geoscience major, the faculty advisor should discuss the curriculum, including concepts, skills and competency expectations for graduates, as well as detail those that align best to support their aspirations for graduate school or employment. Advisors and students need to also discuss the culture of the geoscience department, the professional societies, as well as extracurricular and social networking opportunities that support overall success.

As previously outlined, quantitative skills are a strong predictor for post-college success, yet are also viewed as a key recruitment and retention barrier. Many departments have piloted approaches to support students in developing their core science and math skills. One successful approach has been to integrate math, chemistry, physics, and computational science into all levels of geoscience courses so the students can both learn these topics in a geoscience context and understand their application and importance to their degree (e.g., "Math You Need, When You Need It"; see Section 3). Ongoing contextual success builds student's sense of self-efficacy (i.e., belief in their ability to succeed).

To ensure students are building a solid foundation, departmental, program or institutional tutors or pre-calculus and lower level science courses and workshops can address any deficiencies. Rigorous summer bridge programs in mathematics and chemistry can provide for a better transition to college STEM courses (Ashley et al., 2017; Dickerson et al., 2014; Murphy et al., 2010). Some institutions incorporate more pre-college material in lower division science and math courses or teach material at a slower pace. For example, a course may have more contact (and credit) hours or be offered over multi-semesters. Others have tried a modular approach, breaking the material into multiple three-week segments. Some encourage students to take these more difficult courses during the summer when they can concentrate fully on the subject. Other departments offer calculus taught by geoscience faculty or sequence core science courses with contextual geoscience courses, such as one semester of physics and/or chemistry followed by a semester of geophysics and/or geochemistry.

Support is not just about academics; it includes social, economic, and cultural factors (Tinto and Engstrom, 2008). Develop a gathering place where students and faculty can get together on a regular basis to provide community continuity. Student organizations can help with building a sense of belonging, and successful inclusion of diverse students in the academic and social communities of geoscience campuses will address feelings of isolation and exclusion.

As nascent members of the geoscience community, undergraduate students need to be engaged in the development of a career to not only to prepare for their future but to help motivate them to complete their program. Strategies include inviting alumni to talk about what they do in their careers and how they achieved their present position. Having a successful professional geoscientist join a field trip to interact with the students can illustrate context relative to job opportunities. Career information should be provided to students by the department and/or program, and this same information should be shared with your institutional career services office. Students should be encouraged to take advantage of professional networks that have student chapters (e.g., American Association of Petroleum Geologists (AAPG), Society of Exploration Geophysicists (SEG), National Association of Black Geoscientists (NABG), Society for Advancement of Chicanos/Hispanics and Native Americans in Science (SACNAS), American Indian Science and Engineering Society (AISES), the American Water Resources Association (AWRA), etc.). Departments and programs should encourage and financially support undergraduate attendees and/or presenters at regional and national scientific conferences that help to build student self-efficacy as nascent geoscience professionals. Many societies provide scholarships for first time underrepresented minority attendees (AGU, GSA, etc.).

Nurturing a sense of departmental community that provides academic, social, and financial support within the undergraduate program and the classroom as an entrée to the discipline and profession is critical to student success. Departments should accept new students as emerging geoscientists who are part of the field, but starting their learning journey. Field trips are excellent mechanisms to build camaraderie and community that few other disciplines have, though field trip leaders need to be cognizant of issues that may impede participation by underrepresented students.

Effective pedagogy supports retention and student success, especially pedagogies that draw on collaborative and active learning (e.g., *Association of American Colleges and Universities* (AACU) High Impact Practices, Liberal Education and America's Promise (LEAP), Calculus Communities of Scholars (Asera, 2001)) These pedagogies also support inclusive learning environments (Beane et al., 2019) (see Section 5). Other successful strategies include providing explicit instruction in effective study skills, metacognitive instruction to help students understand the way they learn, and more structured opportunities for assessing what they have learned throughout the semester.

Active learning strategies and undergraduate research experiences help students identify their skills and interests more thoroughly, build strong connections to mentors, and understand what future steps are needed to continue in a desired field (e.g., Lopatto, 2007). These activities or research projects may be as simple as data collection or field and lab assistance on a larger project that creates a sense of belonging. Early authentic research experiences for undergraduate students help with retention (e.g., UT Austin Freshman Research Initiative (FRI)[35]; Simmons, 2014; Beckham et al., 2015), particularly for underrepresented minorities. These research experiences can be during the school year or summer, and ideally should be paid if participation could cause a financial hardship by keeping students from jobs. Additionally, summer Research Experiences for Undergraduates (REU) programs sponsored by organizations such as the University Corporation for Atmospheric Research (UCAR), UNAVCO, or the Western Alliance to Expand Student Opportunities (WAESO), the *Louis Stokes Alliances for Minority Participation* (LSAMP) and other NSF sponsored REU programs across the country, are important ways to build student success.

Concern about mental health issues among college and university students has resulted in increased counseling and mentoring resources for all students. Any undergraduate may suffer from feelings of inadequacy, so departments must develop strong mentoring and engagement programs for all students, and especially for underrepresented minority and at-risk students, including students who commute and those living off-campus. Generally, students develop a strong sense of belonging if early in their education they are involved in research, disciplinary projects or contests, and/or student groups or chapters, where they can form working relationships with other students as part of a cohort.

Mental health challenges also impact high-performing high school students who have never had a poor grade. When starting college or the university, they may no longer be the best in the class, or find it difficult to adjust to a new less structured life. If they do poorly on an exam or in a class, they may magnify it out of proportion, be devastated, and unwilling to admit failure to their family. They also are likely to change majors, or because they have never considered the possibility of failure, drop out. They, too, need robust advising and mentoring.

Effective Strategies for Students Underrepresented in the Geosciences

Successful engagement of underrepresented minority students involve several common components: mentoring, active learning, research, and formative

Courtesy of the Jackson School of Geosciences, University of Texas at Austin

35 https://cns.utexas.edu/news/freshman-research-starts-at-ut-changes-the-world

experiences. *These same components are important for all students as well.* Mentoring is a powerful tool that is effective at the peer level, where students already engaged in the geosciences mentor other students; and the faculty level, where students engage with faculty based on aligned interest and skills. Mentoring groups with mixed cohorts of underrepresented and non-underrepresented students have proven to be successful, particularly for freshman. Mentoring is also key in helping individuals interested in geosciences navigate and discover the wide range of careers and opportunities offered.

Another component to underrepresented minority student engagement in the geosciences is a formative field experience. While not true for all underrepresented students, some minorities and other urban students lack formative childhood experiences with outdoor spaces that many geoscientists find intrinsic to the science. For many underrepresented minority students, sponsored field experiences are often the only time they have been introduced to natural science in outdoor spaces (National Park Service (NPS) survey 2008–2009). These experiences can be even more formative for students when they interact with outdoor spaces through the lenses of newly acquired skills and knowledge allowing them to interpret their natural surroundings or an environmental issue. This approach might include connecting underrepresented students to ocean and marine environments through time spent on research vessels and through coastal research opportunities, directly or remotely.

The departments and programs hold the institutional responsibility to provide support networks, safety nets, mentoring, and more (Wolfe and Riggs, 2017) to ensure student success. Success of students underrepresented in geoscience programs is particularly difficult. Departments and programs should continually address the question of why diversity matters, reduce the prevalence of "lonely onlys", and build student self-efficacy. They need to educate faculty and students (particularly underrepresented students) about topics such as "imposter syndrome" where students feel inadequate despite evident success and "stereotype threat" where students feel at risk of conforming to stereotypes about their social group (Spencer et al., 2016). These two conditions can have the largest impact on those students who are doing or want to do well, resulting in decreased performance (Santiago and Einarson, 1998). Geoscience departments must be willing to look inward, change departmental culture, and develop, incorporate, and advertise to their students programs that are welcoming and designed to promote success among underrepresented student groups.

Faculty and counselors need to recognize many underrepresented students do not know anyone else who has gone to college and thus are significantly less likely to have a roadmap for their future. Small setbacks can be amplified out of proportion. Many have deep connections to their family and home, but their families cannot provide the advice and guidance the students need. Robust advising and mentoring are required. Faculty and/or counselors may need to do "intrusive advising", intentionally contacting a student to develop a positive relationship that promotes academic motivation and persistence. Many institutions also have resources to help departments and programs with these issues.

Advisors must consider the home environment of the minority or "first in family" student. A student who commutes to school and does not live on campus leads a different student life than one who lives in a dormitory. A student who is the first in the family or community to attend college does not have a support or information system like that of peers whose families are educated. Students from some cultures must continually manage the expectations of family or community members whose demands can conflict with academic needs. For example, attending a Pow Wow is mandatory and often not scheduled; enthusiasm about school can be viewed as having fun and not being serious about supporting the family; and the lack of understanding of the difference between high school and university work can overwhelm students.

Faculty and staff need professional development on cultural sensitivity and implicit bias, and departments and institutions need to develop robust diversity, equity, and inclusion plans. Many colleges and universities now offer or require such training centrally or through College of Education courses on teaching diverse learners. Some professional societies, including NAGT, offer workshops on these subjects. In the classroom, instructors need to identify and embrace all types of diversity, recognizing there are diverse communities within underrepresented minorities and to learn about the cultures, heritage, skill levels, and learning styles of students you are working with and adjust teaching and mentoring accordingly.

Promising approaches to broadening participation often happen through institutional partnerships among 2YCs, 4-year colleges/universities (4YCs) and minority serving 4YCs (MSI), Hispanic Serving Institutions (HSI), Historically Black Colleges and Universities (HBCU's), and Tribal Colleges and through leveraging research infrastructures and research opportunities to enhance a student academic training. Universities and four-year colleges (4YC) intending to increase diversity need to develop such relationships and collaborations (Box 8.2).

Partnerships between institutions with clearly articulated and communicated pathways for students can additionally provide opportunities for 2YC, MSI, HIS, HBCU and Tribal College students to participate in programs such as REUs, and mentor students before, during, and after transfers to ensure retention and success of high-risk students. Recruitment scholarships for the underrepresented minorities at two-year colleges can be made portable to 4-year colleges if the students transfer.

Box 8.2: Fort Valley State University Cooperative Developmental Energy Program (CDEP)

The Cooperative Developmental Energy Program (CDEP) program[36], founded in 1983 by its Director, Dr. Isaac Crumbly, focuses on the recruitment and placement of academically talented minorities and females into professional level careers in the energy and other STEM-related industries. It has one of the best track-records in the nation for recruiting minorities and women in science and engineering disciplines.

The CDEP dual degree programs have produced 106 engineers, 40 geoscientists, and 9 health physicists. The program achieves its objectives through scholarships, internships, providing career and job opportunities, and dual-degree programs in engineering, geology, geophysics, and health physics. Industry, government, and other universities participate.

The dual STEM degree has students enroll at Fort Valley State University for the first three years and pursue a major in mathematics, chemistry, or biology. For years 4 and 5, students transfer to one of CDEP's partnering universities to complete a major in engineering, geology, geophysics, or health physics. At the completion of the five-year program, students earn a B.S. degree from Fort Valley State University and a B.S. or M.S. from one of CDEP's partnering universities. CDEP's current partnering institutions consist of Fort Valley State University, Georgia Institute of Technology, Pennsylvania State University, University of Arkansas, University of Nevada at Las Vegas, and the University of Texas at Austin.

36 https://www.fvsu.edu/cdep

Courtesy of the Jackson School of Geosciences, University of Texas at Austin

Box 8.3: Community College Partnerships

In 2017 there were 941 public community colleges (two-year institutions) in the U.S., serving 5.8 million students making up 34% of the undergraduate population in the U.S. (McFarland et al., National Center for Education Statistics, 2019). As higher education becomes increasingly more expensive, more undergraduate students are attending community college as a means to save money and graduate with their bachelor's degree on time. With undergraduate introductory courses being the most effective marketing tool to engage undeclared students into majoring in geology, a strong effort must be made to ensure these courses across institution types and modes of delivery are given our utmost attention and support. Community colleges reach a wider market of undergraduate students and therefore should be a key outreach component for departments and institutions as they encourage students to study the geosciences. Many NSF RFPs require that collaborations with 2YCs be incorporated into the scope of work for the proposal and this opportunity should not be overlooked by 4YC or 2YC institutions. The most effective collaborations between 2YC and 4YC are often systematic in their efforts.

Curriculum: Ensure content being taught in a 2YC introductory course prepares students for the higher division courses at the 4YC. Build conversations and formative collaborations between colleagues at each institution which can often lead to research and mentoring projects between institutions that serve to increase 2YC undergraduates' awareness of the 4YC institution. Field research opportunities for faculty and students can be a tremendous incentive for continued collaborations and for matriculation to the 4YC institution after the student graduates or finishes taking 2YC classes.

Degree plans: Many 2YCs have 2-year Associate degrees in Science with some even specific to the Geosciences. Finding out if a 2YC has a degree that is transferable, and ensuring that the courses within the degree are transferable and are equivalent to a student who has the same course work and hours at the 4YC, will help ensure that students do not use up valuable educational resources (financial aid, college credits, time, money, etc.)

Mentoring: Student mentoring stretching between 2YC and 4YC greatly improves the success of students making the transition from the community college system to the university system. Having a faculty mentor at both institutions can help facilitate this transition and the challenges that arise such as larger class sizes and increased cost. Mentors can help students navigate these hurdles and increase the chance a student continues and does not drop out. Mentors can be faculty mentors, departmental advisors, and even departmental undergraduate or graduate students.

2YCS OPENING THE DOORS TO THEIR STUDENTS' FUTURE

A critical gateway to engaging the broader population is through 2YC (community college) institutions (Box 8.3). Community colleges often have a high percentage of underrepresented minority students and as the cost of higher education increases, many non-minority students are enrolling as well. Building community college partnerships to bridge the 2- and 4-year college divide provides multiple opportunities to introduce underrepresented students to geoscience at critical decision points.

OBSERVATIONS FROM THE COMMUNITY:

Developing relationships is critical. We recently had a weekend field trip and invited a faculty member from the transfer university to attend. Having that unstructured time for majors to ask her questions was invaluable and was something I hope to continue to accomplish and work toward in the future. *(2-year community college)*

Robust communication between advisors and faculty at local 2YCs and 4YCs and joint advising strategies are beneficial for successful student transfers. Advising and mentoring needs to happen before, after and during the transfer process. These relationships lead to increasing 4YC enrollments, potentially increased diversity, and a greater chance of long-term student success as preparation and expectations for transfers will be closely aligned to the target 4YC.

Early and intentional integration of transfer students from 2YC to 4YC settings is important for ensuring their success, retention, and graduation. Students struggle to feel a part of their new program when they transfer in with half of their degree program already completed, particularly in into programs with an established strong community. The 4YC students will have already developed cohorts and friendships. Activities that integrate transfer students into the 4YC community before and after the transfer build community and provide support. Joint mentoring of students by 2YC and 4YC advisors throughout the process helps ensure retention and success of high-risk students. Successful approaches have included joint fieldtrips for freshmen and sophomores at the two institutions, summer research internships, REUs, or field programs for 2YC students at 4YCs, and collaboration with upper classmen at 4YCs (Boxes 8.3, 8.4). Relationships can be strengthened by 4YC faculty inviting 2YC students and faculty to research talks and symposiums, co-advising students, and establishing peer and vertical mentoring programs between the two institutions. These partnerships offer opportunities to a broader segment of the student population at both the 2YC and 4YCs than just the transferring students. Institutions need to be attentive to managing costs, such as those associated with funding to house and support students during REUs.

Respondents in the *2014–2015 survey* indicated limited interaction between 2YC and 4YC (Fig. 8-2). About 17% of participating Heads and Chairs at the *2016 Summit* and subsequent workshops that submitted a progress report said they successfully implemented some of these 2YC–4YC engagement strategies. These included joint fieldtrips, student panels, social events, development of a transfer pathway academic map, and increased interactions between faculty and students at local community colleges and 4-year colleges/universities to ease success of transferring students (Box 8.4).

To prepare students for future success with their educational pathway, 2YC faculty should teach students to be good learners who are open to new ideas, engage them in problem solving, and develop team skills; discover and use any non-traditional skills they already have; and leverage local professional societies for both 2YC students and faculty. Faculty at 2YCs need to be active in the professional community and participating in national 2YC faculty networks, bringing benefits to themselves and their students. Informing administrators about professional activities is important, so publicize good-news stories about engagement with outside stakeholders and professional activities.

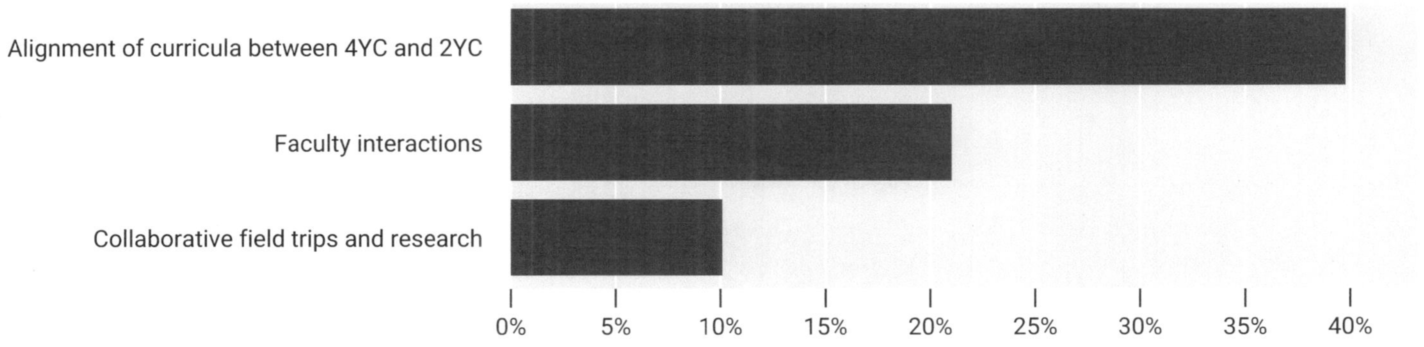

Figure 8-2: Methods Used to Ease 2YC-4YC Transition
Departments using specific methods
Survey Question: Which of the following does your department do to ease the transition between 2-year and 4-year colleges?

Courtesy of Joshua Villalobos

Box 8.4: University and Community College Collaboration

The University of Texas at El Paso (UTEP) and El Paso Community College (EPCC) have formed a strong collaborations over many years that incorporates a partnership between their geoscience faculty to ensure that EPCC students interested in the geosciences have a high success in matriculation to, and graduation from, UTEP. Geoscience faculty at both institutions convene regularly to discuss any recent changes or recommendations to either curriculum objectives or degree plans. This ensures EPCC students who major in Geosciences are receiving the required instruction and transferable credit for them to succeed at both at UTEP and EPCC. Collaborations between faculty and students at UTEP and EPCC are often strong components in grant proposals submitted by UTEP and incorporate EPCC students facilitating in various field and laboratory tasks that are often considered routine (assisting if deploying field equipment, running lab equipment, data input, etc.) but in fact can be very formative to a student who has begun to show an interest in the geosciences and who may not have yet been on a university campus. These activities have proven to be highly effective in increasing UTEP's undergraduate programs and increasing the number of geoscience majors at EPCC. Faculty at both institutions have acquired enough knowledge about each other and their respective institutions that they have begun to serve as successful mentors at both ends to assist students in the challenges many face transitioning between community college and the university.

One effective collaborative project is SLATES (Service Learning Activities Targeting the Earth Sciences) that aims to diversify service learning opportunities for undergraduates at the Hispanic Serving Institution, UTEP, and EPCC. A series of short-term activities (< 10 hrs./semester) were developed to target students in introductory geoscience courses to help increase the number of geoscience majors, as well as long-term (>10 hrs./semester) activities for majors to apply their knowledge and skills outside the classroom. In the first year of SLATES, we focused primarily on short-term activities while laying the groundwork for longer-term activities and encouraged students to assist in designing new activities. EPCC students chose to focus on water and sustainability projects, including serving as tour guides for the El Paso Water Utilities' TecH_2O Center and desalination facility, and developing lesson plans on El Paso's groundwater for 4th grade students. These activities involved over 170 students (K–14), parents, and teachers. Students at UTEP focused on tutoring lower division majors in key classes (e.g., mineralogy, petrology) and K–12 outreach (involving over 40 students). We also organized a variety of field trips that highlighted local geology for students, family, and friends. A total of 242 participants (35% non-science majors) attended or assisted with the local field trips.

Students at 2YCs generally mirror the community demographic and can thrive at 4YCs, especially when appropriate support structures are in place. 4YC Faculty working with transfer students from 2YCs need to overcome any cultural or academic biases — attendance at a 2YC or underrepresentation does not mean underprepared. 2YC students are fundamentally the same as 4YC students, and 27% 4YC students completed two years at a 2YC before finishing their last two years at 4YCs (also 26% of M.S. and 17% of Ph.D.'s) (Wilson, 2018).

For success, a clear pathway to degree completion for transfer students must be defined and communicated to local 2YCs to ensure students are on track at the end of their first two years to be prepared for the next two. Faculty at both institutions need to coordinate course objectives, curriculum, and degree plans. They should discuss content, objectives, and any evolution or changes in degree programs. Clear articulation agreements are needed, but that goes beyond course numbers; it is critical in advising students to ensure transfer courses *actually* transfer as specific courses needed for the degree, not just as credit hours. If 4YCs have transfer students from multiple 2YCs, the 4YC faculty should work with the network of 2YCs. Also, if a student transfers before receiving their Associates degree from the 2YC, it is important for faculty at both institutions to facilitate cross-transfer credits from the 4YC to the 2YC so the students can still be awarded their associates degree. This coordination will ensure that students who do not finish the four-year degree will at least have the Associates degree and increases the likelihood that they complete the four-year degree. Also, 2YCs are evaluated on the number of Associate degrees that are completed.

BROADENING PARTICIPATION AND INSTITUTIONAL CHANGE

Changing the culture of academic departments through transformative institutional practices to sustain diversity efforts progresses slowly. Problems and solutions do not lie with the communities we hope to serve but are the responsibility of leadership, who are in position to motivate change and re-think what constitutes a geoscience degree. Building on research and evaluation in best practices, academic programs are beginning to embrace new approaches to increasing the diversity of majors, career options, and training. A snapshot of current geosciences diversity programs funded by the NSF shows innovation in approaches and skill development around topics that are particularly meaningful to students, urban or rural. Researchers should take advantage of the Broader Impact review criterion for federal grants to continue to initiate and test new programs and ideas to address minority and nontraditional student issues in the geosciences.

ROLE OF PROFESSIONAL SOCIETIES (GSA, AGU, NABG, SACNAS, AISES, NAGT)

Geoscience and other STEM professional societies have been at the forefront for decades in building strategic approaches for recruiting underrepresented populations into the sciences. These initiatives have largely been structured as competitive scholarships or overarching informational or mentoring initiatives. Efforts by individual programs, often funded by the National Science Foundation and industry, have been key to capturing these newly engaged students. However, this process

OBSERVATIONS FROM THE COMMUNITY:

Institutionally our students (not just geology but STEM and non-STEM) are receiving their Associates' through reverse articulation. (That is) wonderful for the student to achieve this intermediate-step by earning a degree) and very good for our 2YC institution to have this "degree awarded" data points for our state legislative funding formula (a formula that does not recognize the course transfer mission as a fundable merit). *(2-year community college)*

I convened two summits with area 2YC instructors, revise our 2+2 transfer plans, and disseminated them to the appropriate college recruitment offices and 2YCs. *(Doctoral-granting university)*

We developed a freshman field exchange with the local community college where the two groups of students did local fieldtrips together. *(Doctoral-granting R1 university)*

struggles to scale effectively, as demonstrated by the marginal improvement in participation by underrepresented populations. Additionally, efforts by the National Academy of Engineering and the American Geosciences Institute demonstrated the critical barrier to success has been building sufficient self-efficacy in these recruited populations (Houlton and Keane, 2017). Starting in 2020, a spectrum of major NSF-funded initiatives have been launched with professional societies to develop community-wide networks that foment change to improve the diversity of the geoscience community.

Vincent Hoban, UCD Media Services

Recommendations:

- Address the public perception of the geosciences by emphasizing societal, economic, and employment relevance
- Emulate and develop program/department-focused positive recruitment programs for new students, lower division non-majors, transfer students, and students underrepresented in the geosciences, taking advantage of institutional efforts
- Develop or collaborate with STEM programs for minority students at pre-high school and high school levels, as appropriate for your institution
- Develop a formal approach for student retention and success that includes mentoring, building a sense of community, and other supportive aspects, particularly focusing on students underrepresented in the geosciences
- Develop programs that facilitate the success of transfer students from 2YC to 4YC and/or universities before, during, and after transfer
- Build institutional partnerships among two-year colleges, four-year colleges/universities and minority serving institutions (MSI) and leverage research infrastructures and research opportunities to enhance a student academic training at Hispanic Serving Institutions (HSI), Historically Black Colleges and Universities (HBCU's), and Tribal Colleges, leveraging institutional efforts as appropriate
- Initiate and test new programs to address underrepresented minority and nontraditional student issues in the geosciences
- Use the Broader Impact review criterion for federal grants to encourage actions to increase underrepresented minority participation, retention, and success

9. Leveraging the Next Generation Science Standards in Introductory and Non-Major Courses to Recruit Majors and Prepare K–12 Science or Geoscience Teachers

Courtesy of the Jackson School of Geosciences, University of Texas at Austin

Introductory and non-major courses should leverage the Next Generation Science Standards to engage all students and preservice teachers in the geosciences.

CHALLENGES AND OPPORTUNITIES

The Next Generation Science Standards (NGSS)[37] are fueling a major reform in K–12 science education in the U.S. and offer an unprecedented opportunity to expand the reach of the geosciences (Wysession, 2014). Introductory and non-majors courses can leverage the NGSS to better engage all students and support pre-service teachers in the geosciences. As of 2020, 20 states have adopted the NGSS and 24 others have developed standards based on the Framework for K–12 Science Education[38]. In using these standards, K–12 education focuses on active-learning pedagogies, student-centric education, cross-disciplinary science, and key disciplinary concepts. Within this approach, the geosciences are well-positioned to serve as a vehicle for addressing the requirements for integration of the science disciplines while simultaneously incorporating mathematics, engineering, communication, and societal considerations (see Box 9.1 for specific information). Although most states do not require students to take an Earth science course in middle or high school, the NGSS elevate the importance of Earth and space science in pre-college science education to a level equivalent to that of the life and physical (physics and chemistry) sciences.

As NGSS adoption expands, students will enter college with different expectations for how science works and is taught. We need to prepare for these students by becoming familiar with NGSS and adapting our undergraduate programs. If we use the recommendations discussed in previous sections regarding active learning pedagogies (Section 5), and a focus on concepts, skills, and competencies (Section 3), our courses will align with the expectations of the NGSS and these future students. Reforming introductory and non-major geoscience classes is essential for all students, including geoscience majors and pre-service K–12 science and geoscience teachers (e.g., PCAST, 2012), because it will contribute to improving geoscience literacy for all. Effective pedagogues, including societally relevant, active-learning opportunities, in introductory courses may entice more students to major in the geosciences. Departments should provide faculty teaching introductory and non-major

37 https://www.nextgenscience.org/

38 see https://ngss.nsta.org/about.aspx

Box 9.1: Pre-college Science Education in the U.S.

K–12 science education in the U.S. is currently undergoing major reform that has produced the Next Generation Science Standards (NGSS). For the Earth and space sciences, this reform began with a series of community-based collaborations that produced the "Blueprint for Change: Report from the National Conference on the Revolution in Earth and Space Science Education" (Barstow et al., 2002) followed by a series of materials (UCAR, 2007; ESLI, 2010; OLN, 2015; EERE, 2017; CLEAN, 2018) designed to promote understanding of the "big ideas" in the geosciences. These documents, among others for other disciplines, guided the development and release of the Next Generation Science Standards (NGSS; NGSS Lead States, 2013a) that continued reframing science disciplines that started with National Science Education Standards in 1996.

The three-dimensional nature of the NGSS is truly revolutionary. The standards are articulated as a series of "performance expectations" that include a component drawn from each of the three dimensions (that is, they each include a practice, a crosscutting concept, and a disciplinary core idea). The performance expectations specify what a student should be able to do in order to demonstrate mastery of all three dimensions of the standard. Evidence statements (NGSS Lead States, 2013c) for the NGSS provide teachers and administrators with guidance regarding what should count as credible evidence that a student has mastered a standard. Comprehensive three-dimensional assessments address all components of the target performance expectation(s) simultaneously. A series of comprehensive assessment examples addressing multiple performance expectations is available online from the NGSS website (NGSS Lead States, 2013b). The standards development was based on decades of research on effective practices in science education, much of which was summarized in the National Research Council's publication "A Framework for K–12 Science Education: Practices, Crosscutting Concepts, and Core Ideas" (NRC, 2012).

classes with opportunities for professional development related to implementing NGSS guidance and strategies so that they can learn about the research-based strategies that are being implemented in K–12 classrooms.

Preparing geoscience-literate K–12 teachers will help the geosciences to address critical workforce needs while also promoting development of a geoscience-literate citizenry. Some of the most pressing problems facing the world are related to the geosciences, and it is crucial that the public, most of which will not pursue a post-secondary science credential, has a basic understanding of the Earth and its systems. The *2014 and 2016 Summit* participants overwhelmingly agreed that our best approach is to train K–12 educators, from any disciplinary background, to integrate geosciences into their classes in accordance with the NGSS. The participants' recommendations for developing the next generation of introductory courses are outlined below. K–12 teachers who have taken effective geoscience courses in college are better equipped to educate their students about how the world works, demonstrate the relevance of this knowledge to local and global issues, and instill enthusiasm and interest in the geosciences.

A cultural shift to focusing on the role of geoscience programs in influencing K–12 teacher preparation is needed. Of all the sciences, Earth science has the fewest disciplinarily trained K–12 teachers (Wilson, 2019). We need to expand efforts to prepare geoscience-literate K–12 teachers to increase pre-college exposure to geosciences. By helping teachers use authentic geoscience examples to promote science literacy, geoscience departments can strengthen public understanding and promote recruitment of future majors.

REDESIGN OF INTRODUCTORY AND NON-MAJOR COURSES

Faculty teaching introductory and non-major courses should become familiar with the structure and expectations of NGSS. The NGSS has three core disciplinary domains: Physical Science, Life Science, and Earth and Space Science; Engineering is treated at the same level as Science. For the first time, Earth and Space Science is considered at the same level as other sciences. Instead of emphasizing disciplinary content above all other considerations, the NGSS pay equal attention to three different dimensions of science:

1. practices of science and engineering;
2. crosscutting concepts that span and unite all of science; and
3. disciplinary core ideas.

This "three-dimensional" nature of the NGSS is truly revolutionary. As students proceed through the learning process, they employ the scientific practices and use the crosscutting concepts to develop a deep understanding of the disciplinary core ideas. They demonstrate mastery of a standard by using all three dimensions of science to investigate a phenomenon or develop a solution to a problem. The NGSS were developed based on evidence of what works, and through an extensive community-based process. The recommendations for geoscience majors as discussed in previous sections (Sections 3, 5 and 8) are fully aligned with this process.

The NGSS call for implementation of student-centered pedagogies that empower students to drive their own learning. Teaching with real data and models can support student-centered pedagogy while also promoting a deep understanding of material (e.g., Freeman et al., 2014; NRC, 2015; Miller & Kastens, 2018; Kastens et al., 2019).

To increase the relevance of introductory geoscience courses to all students, majors and non-majors alike, faculty should focus on phenomena and problems. An earth-systems approach, which emphasizes the integration of physics, chemistry, biology, math, societal implications, and communication will help increase students' perception of the relevance of the geosciences. By employing active-learning strategies, instructors can help their students develop an awareness that the Earth acts as a complex system and to recognize the nature of the linkages between different parts of the system. The students should have opportunities to explore the causes, effects, and feedback loops within the Earth system, and investigate how geoscience processes impact humans and how humans impact the Earth systems (Ireton et al., 1996).

PREPARING AND SUPPORTING K–12 TEACHERS FOR SUCCESS

Most K–12 and Earth science teachers only take introductory or non-major geoscience courses (Banilower et al., 2013; Gilbert et al., 2019); thus redesigning these courses to incorporate the NGSS framework and pedagogy is imperative to preparing pre-service teachers (Egger, 2019). Integrating the vision of the NGSS into university curricula using the geoscience literacy documents[39] will strengthen the curriculum and improve the preparation of future teachers.

Teachers' preparation will be enhanced if geoscience-related phenomena and problems are included in introductory and non-major courses in ways that can be modified for implementation at the K–12 level, whether in K–12 Earth and space science courses or as geoscience examples in other disciplines' courses, such as physics, chemistry, biology, and math. The introductory geoscience courses need to model the best pedagogy and investigation of scientific phenomena and problems that can be adapted by future teachers. Such courses will prepare future science teachers to use the NGSS and real-world examples in their K–12 classrooms.

Pre-service teachers who major or minor in the geosciences need to engage in authentic geoscience experiences (i.e., field trips, service learning, laboratory-based projects or work experiences) along with rigorous disciplinary preparation. They also need methods courses that emphasize strategies and pedagogies validated by education research, designed in collaboration with education and/or geoscience education faculty. Pre-service teachers who have the opportunity to do research and get early experiences in K–12 classrooms while completing their undergraduate degree will be well-prepared to introduce their future students to the geosciences.

K–12 teachers, who are already in the classroom, will benefit from efforts undertaken by the geoscience community to develop low cost resources, including laboratory, field, and problem-solving exercises that can be easily adapted for high school classes. Useful teaching materials and methods are crucial to increasing geoscience content in K–12 classes. The best repository of tested undergraduate-level teaching resources resides on the Science

39 http://www.earthscienceliteracy.org/

Education Resource Center (SERC) website accessible through NAGT's "Teach the Earth" portal[40]. Other excellent resources are available through geoscience consortia and organizations such as IRIS, UNAVCO, and AGI, and government agencies including NOAA, and NASA, and the U.S. Department of Agriculture Natural Resources Conservation Service Web Soil Survey that can be used anywhere in the U.S..

Geoscientists, and especially faculty, need to value K–12 teaching as a career. Education majors and pre-service Earth science teachers are as important as geoscience majors and are critical to developing the next generation of geoscientists and Earth literate citizens. In all classrooms, there are a diversity of skills and trajectories and each is important to develop.

K–12 TEACHER EDUCATION PROGRAMS

For 2YC and 4YC geoscience programs already invested in K–12 teacher education, contributing to K–12 education reform is necessary. Even a cursory review of the NGSS performance expectations and evidence statements makes it clear that teachers can no longer rely on worksheets or multiple-choice tests to gauge student learning. Thus teachers, to be effective and prepare their students for success on state-based high-stakes tests, must learn how to develop and score three-dimensional assessments. These new types of assessments typically ask students to demonstrate mastery in ways that differ significantly from traditional pencil-and-paper classroom-based exams. Although guidance exists for developing three-dimensional performance-based assessments (NRC, 2014), pre-service teacher education programs must increase the focus on this activity because it is challenging for both new and experienced teachers. Programs must give teachers time to practice and reflect on the nature of assessments to prepare them for success in the classroom. These assessment strategies will also contribute to improved assessment of student learning that can be adapted for undergraduate classes as well.

One of the most significant aspects of the vision underpinning the NGSS is the transition from a traditional teacher-centered classroom to student-centered classrooms in which the teacher serves as a guide on the side rather than taking on the role of a sage on the stage in the traditional classroom model. In a fully student-centered classroom, the students ask the questions and develop the answers to those questions. The teacher keeps the learning process moving forward and helps to provide resources as needed, but does not serve as the ultimate source of knowledge.

The transition to student-centered pedagogy requires implementation of NGSS-aligned curricula by teachers who are prepared to implement NGSS-aligned pedagogy (NRC, 2015). Novice teachers will find effectively implementing NGSS-aligned instruction challenging if they have not had the opportunity to experience it as part of their education (NASEM, 2015). It will also be difficult for experienced teachers unless they have intensive and sustained professional learning involving substantial hands-on practice. Thus, effective implementation of the NGSS and other Framework-aligned standards will require more, and a different type of, pre-service and in-service teacher education. Geoscience faculty, particularly those who work with pre- or in-service teachers, need to be prepared to implement student-centered teaching and the three-dimensional approach to learning in their own classrooms.

To develop proficiency with student-centered teaching methods and the NGSS, geoscience faculty involved in K–12 teacher education should engage with the colleges/schools of education and effective K–12 teachers to understand how they are addressing NGSS. Collaboration with two-year college faculty who teach a broad spectrum of students can support development of culturally relevant practices. The broader the engagement of geoscientists with pre-service and in-service teacher professional development, the larger the potential for K–12 students exposure to geoscience.

The *2014–2015 survey* shows institutions beginning to integrate the 2013 NGSS into curricula and 51% integrating math and basic sciences into undergraduate courses (Fig. 9-1). In addition, about 30% of the respondents indicated that their institution is involved in K–12 teacher preparation and 36% offer professional development programs for in-service K–12 teachers (Fig. 9-1; Appendix A). Very few participating Heads and Chairs at the *2016 Summit* and subsequent workshops that submitted a progress report, however, indicated any changes in in K–12 teacher preparation.

For faculty working with pre- and in-service teachers it is important to:

- be a good geoscience role model and mentor;
- demonstrate respect for the role of K–12 teachers as academic professionals;

40 https://serc.carleton.edu/teachearth/

- be willing to discuss different pedagogies pros and cons;
- connect to the NGSS whenever possible;
- devise strategies to address and assess multiple dimensions of the NGSS simultaneously;
- continually inform local teachers about new Earth science resources, including locally relevant project;
- and develop local field guidebooks.

Departments should also consider more involvement with in-service teachers and engage with teacher networks in their area. Seek out and work with existing professional development programs for in-service teachers, or if none exist in your area, use existing resources to develop one.

Pre-service teachers, like all students, need career advice, information about job opportunities, and mentors. Connecting pre- and in-service teachers to professional organizations (GSA, AGU, NAGT, etc.) will help them develop professional networks. Departments should foster personal connections among faculty and pre-and in-service K–12 teachers and continue to build those long-term collaborative relationships as the teachers enter the classroom. Collaboration with teachers may involve science or educational research, professional development, outreach, or development of standards-based teaching resources. Many possibilities exist, including using in-service teacher's expertise to design non-major and introductory undergraduate classroom and laboratory activities, or developing pilot programs combining geoscience content with content from other fields for use in high-school, introductory, and non-major geoscience courses.

When collaborating with in-service teachers, it is important to recognize their expertise as educators. Pre- and in-service teachers should be included as co-authors on scholarly works, supported to attend state, regional, and national conferences, and encouraged to present at those events. Encouraging interaction between undergraduate and graduate students and K–12 teachers, including K–12 classroom involvement, is valuable in expanding the teachers' network of geoscientists, improving graduate and undergraduate students' communication skills, and providing the K–12 students with geoscience role models. The NSF Graduate STEM Fellows in K–12 Education program was successful in this effort, and projects supported through that program provide good models for success (Ufnar et al., 2012).

By inviting K–12 teachers to participate in activities, including attending and/or

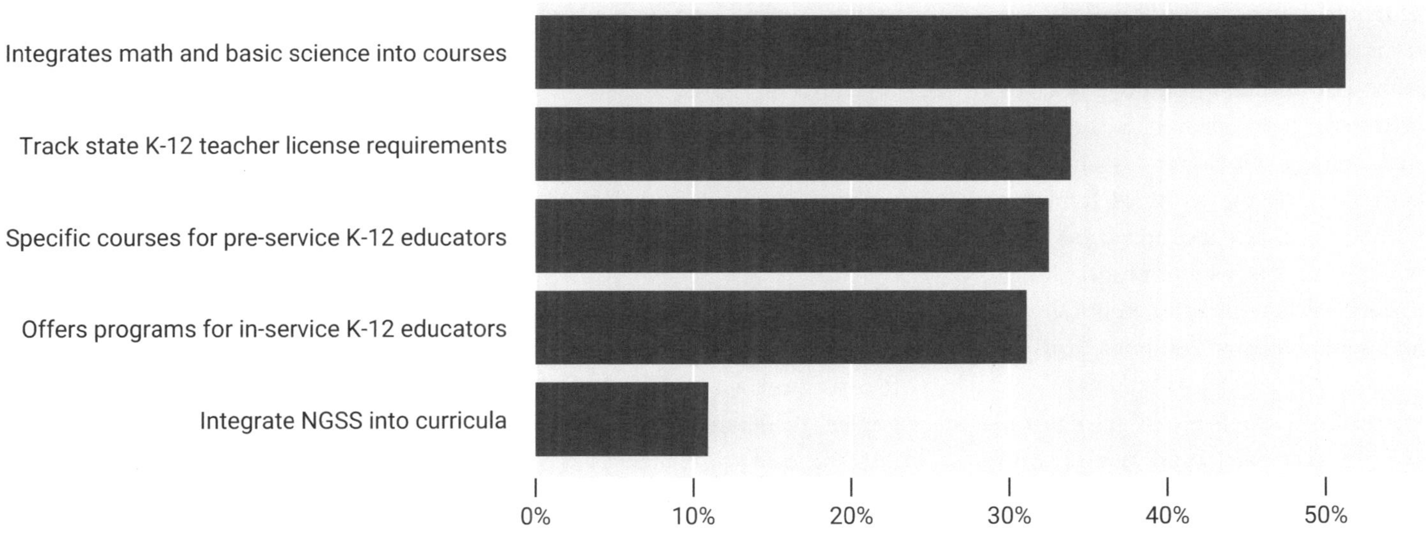

Figure 9-1: Departmental Supports for K-12 Educators

Departments offering specific supports for K-12 educatiors

Survey Question: Which of the following does your department do to help with preparation of K-12 teachers?

presenting seminar talks, participating in field trips, serving on advisory boards, and other activities, post-secondary institutions can contribute to teachers' professional development while also promoting development of sustainable collaborations among teachers, faculty and students.

If teacher education is an important aspect of an undergraduate geoscience program, departments should provide incentives for faculty to participate and include such activities in the reward structure.

In addition to being familiar with NGSS, key faculty members and staff involved in training teachers should also be aware of state education and licensure requirements and state expectations for K–12 instruction and assessment of students in the content areas. The available and required training, individual state standards, and certification requirements vary widely by state, and sometimes even by school district. It is also critical to engage with local school administrators, local school boards and state education boards as changes to K–12 curriculum and requirements can have a major impact on your future students. "In order to fully realize a diverse and well-prepared K–12 Earth and Space Science teacher workforce, teacher education research must also recognize the complex landscape in which teacher education takes place, involving an interplay of programmatic, institutional, demographic, political, state, and national factors" (St. John, 2018).

Few teachers teach Earth science full-time, driven by partial allocations because of limited discipline awareness and licensing limitations. Many formally-educated Earth science educators teach other topics, just as many of the teachers of Earth science come from other disciplinary backgrounds, such as biology. To increase the number of Earth science teachers to meet increasing demand because of the elevation of geosciences in NGSS, institutions of higher education should consider offering a master's in Earth science education (non-thesis), having separate labs and/or classes focusing on how to teach geoscience, and offering field courses designed for pre- and in-service teachers.

IMPLICATIONS OF K–12 SCIENCE EDUCATION REFORM FOR UNDERGRADUATE GEOSCIENCE EDUCATION

As student-centered, three-dimensional science education becomes the norm at the K–12 level, students entering post-secondary institutions are likely to have new expectations for undergraduate education. Undergraduate geoscience programs will need to respond accordingly to remain relevant and interesting as the expectations of students evolve. Also relevant to undergraduate education is conducting research on whether student-centered courses that provide practice with all three dimensions of science are effective in achieving valuable intended learning outcomes.

Recommendations:

- Revise introductory and non-major courses using the NGSS approach focusing on active learning pedagogies, problem- or project-based learning, cross-disciplinary concepts, emphasis of disciplinary core ideas, and investigation of phenomena
- Provide and take opportunities for geoscience faculty to participate in professional learning experiences that introduces them to the nature of the three-dimensional NGSS and associated pedagogies
- Prepare geoscience majors, non-majors and future K–12 science and Earth science teachers by focusing on processes and systems, integrating content from other sciences and math in introductory and non-major courses, using active learning and student-centered pedagogies, and developing and using resources that can be easily adapted for use in middle and high schools
- For pre-service teachers, use and explicitly identify geoscience examples that connect to other sciences, math, social sciences, and communication. Take into consideration that an introductory geoscience may be a student's only science course and that future teachers will be more likely to use geoscience examples if they also support learning of other subjects
- Develop a focus on K–12 teacher education if appropriate to your institution. Include courses that provide practical applications of strategies informed by the literature on geoscience education and other education research and offer training on the implementation of NGSS-aligned teaching resources

10. Building a Learning Continuum for Life — Undergraduate Geoscience Education as Part of the Continuum of Career and Lifelong Learning

Programs and students must recognize that formal undergraduate education is a robust foundation for lifelong learning in support of a successful career

As geoscience work increasingly demands a broad portfolio of knowledge, technical abilities, and professional skills in areas that are rapidly changing, an individual can no longer rely on their formal education and work experience to stay relevant. Students' undergraduate education generally focuses on core geoscience knowledge and skills and the requisite educational outcomes of a balanced undergraduate education. It forms the foundation for lifelong learning and facilitates further training and education not encompassed in undergraduate geoscience program. The traditional 120 credit limit for undergraduate degrees is insufficient to also educate students in professional skills, such as communications, economics, and project management, let alone the specialty skills, advance knowledge, and regulatory understanding required for workplace success. Even a post-graduate master's experience may not fully prepare students for all the skills needed, (e.g., see skills needed for Environmental Consulting; Box 10.1).

Box 10.1: Selected Skills Needed for Geoscientists in Environmental Consulting

- Core geoscience skills
- Chemistry
- Physics
- Biology
- Environmental Engineering
- Soil Science
- Hydrogeology
- Hydrology
- Field mapping/data collection/instrumentation (e.g., LIDAR, GPR, etc)
- GIS/spatial data analysis and management
- Drone-based acquisition (emerging)
- First Aid/CPR/AED training
- OSHA HAZWOPER training
- Technical writing and speaking
- Business writing and speaking
- Business finance
- Environmental compliance/regulations/law course
- Scientific and business ethics

Compiled from Houlton, 2015; Wilson, 2018; *2015 Geoscience Employer Workshop*

The importance of knowing how to learn was stressed at both the *2015 Undergraduate* and *2018 Graduate Geoscience Employers Workshops*. Students need to recognize their formal education is only one component of the learning they will need while in school and during their career. One outcome from any undergraduate degree program is that students should learn how to learn and should become intentional learners as they focus on building specific knowledge and skills during their career. A portion of the needed skills will be developed through their formal education experience, but many others will need to be built either through co-curricular activities or ongoing professional development.

For students to continue to learn, undergraduate degrees must provide them with the foundations for a learning continuum, including core knowledge and principles, and prepare them for geoscientific and systems thinking, while proactively learning and applying new skills and knowledge. This preparation is critical for those starting careers immediately after graduation and for those going on for master's and doctoral degrees.

EXTERNAL CERTIFICATIONS

Geoscience programs and their students should be aware of the opportunities and requirements for professional geoscience licensure. While many professions have overarching accreditation agencies, such as the Accreditation Board for Engineering and Technology (ABET) for engineers, geoscience does not. Nonetheless, with 32 states requiring a license for a geologist to practice professionally (with some employment exceptions), it is critical for students planning a career as a geoscientist to understand the licensing process within their state. The National Association of State Boards of Geologists (ASBOG®) administers two exams: the Fundamentals of Geology exam is taken either during the final semester of college or shortly after graduation; and the Practice of Geology exam is taken after passing the first exam and having completed a specified amount of required work experience for licensure. Globally, licensure of geologists is even more common than in the United States and represents an opportunity for employment mobility.

Professional licenses require ongoing continuing professional education. However, accumulation of Continuing Education Units (CEU) (also often referred to as Professional Development Hours — PDH) is not limited to those who have completed their formal studies. As an extracurricular activity, students can take in-person and online professional development courses that provide recognized CEUs. The American Institute of Professional Geologists (AIPG), in cooperation with AGI, offers on-demand, online professional development courses with a nominal fee for all students to gain CEUs and begin this professional development and certification process.

Through professional development courses, students can build and document their professional skills with coverage of critical topics such as ethics and regulatory compliance, as well as exposure to technical topics in their field of interest so they gain an understanding of how the science is used in an applied context.

LEARNING TO LEARN FOR LIFE

Rapid advancements in technology and science requires students and professionals to master the fundamentals, evolve their understanding, continue to learn new concepts and skills, and even change their view on "settled science." Just as most professional licensure programs require certain levels of ongoing education to retain currency, all professionals need to continue learning over the duration of their careers.

One fundamental transition for most students is moving from the formal education environment to the approaches used within professional and continuing education. Professional development experiences commonly use different modes of delivery compared to formal education, such as on-demand online courses, intensive short courses, or developmental workshops.

The scope and approach of continuing education experiences is varied, whether hyper-focused on a specific technical application or looking at the development of a professional skill within an organizational or disciplinary context. Additionally, the mode of communication is often closer to guiding tutorials, with an expectation that the learner be either versed in, or developing, the habits of mind of the profession (Coble, 2019). The continuing education environment relies on teaching an intentional learner, which can be a substantial change in posture for a student transitioning from a formal education environment.

Some skills desired by employers may be difficult to get in the regular course of study within the existing curriculum — such as computer programming or data analytics. In a reformed curriculum, these skills may well be embedded to help students meet the needs of the modern workforce, but there are substantial external resources, such as MOOCs, code academies, etc., where students can gain some level of certification proving their exposure to these skills. Facilitating the immediate utility of a continuing education experience is critically important, especially if it can be coordinated within a degree program. Developing well-defined skills and competencies is a prime outcome of an intentional learning approach and enabling students to utilize these skills in their courses builds competency and efficacy.

THE CHANGING WORKFORCE

Three factors are changing the nature of work in the geosciences and the expectation of what a geoscientist is: generational turnover, automation of "middle-skills," and evolution in how geoscientists work. All three factors are coming into play in 2020 and will likely define the careers of a generation of students.

First, the geosciences have been preparing for over a decade for the "Great Crew Change" as Baby Boomers retire (Martinsen et al., 2012). This generational shift has already occurred in the minerals industry and is well underway in the energy sector. The environmental sector does not face quite such a generational shift, but it faces the retirement of many of its most experienced workers. This generational shift, when coupled with technological advances, is leading to the elimination of many traditional geoscience occupations, and current recent graduates are defining new geoscience occupations that meet the needs of today and into the future (Malchuk, 2018).

Second, the advancement of data analytics and machine learning is transforming the nature of work in the geosciences (Keane and Wilson, 2018). In 2017, the mining industry analyzed how their geoscientists spent their time and found that 80–82% was spent collecting, collating, and preparing data for analysis. The petroleum industry had similar findings during a 2018 study, with their geoscientists spending 79% of their time wrangling data (Malchuk, 2018; Salamis, 2017), i.e., cleaning, restructuring, and processing raw data into a more usable format. Many companies are using advanced data analytics and machine learning to automate a substantial amount of the data wrangling and basic technical evaluation previously done by those geoscientists, freeing more time for the geoscientists to apply their domain knowledge to increasingly complex problems. Though this automation is beginning to displace middle-skill geoscience work such as routine core logging, stratigraphic interpretation, and data discovery, it is freeing up geologic talent to tackle intellectually more challenging problems and fundamentally increasing the value of the geoscientist to the employer (Keane and Wilson, 2018).

Third, as in many complex fields, collaborative work is critical to success. For the geosciences, there has been a natural progression as the interdisciplinary nature of the science has become integral to its definition. Having evolved from subdisciplines, to collaborations, to the current standard of integrated teams of specialists, the emerging trend with major employers is towards teams of integrated geoscientists who all possess a base level of broad skills in the pertinent areas of geoscience, engineering, and business issues, but with each member bringing particular strengths.

These factors result in the expectation that the future geoscientist is a professional with a broad spectrum of competency and selected areas of mastery (Fig. 10-1). This trend was reflected during both the *2015 Undergraduate* and *2018 Graduate Geoscience Employers Workshops* in which employers emphasized that new employees need to be well-versed across all the major geoscience concepts and conceptually literate of the ideas within them, while also having the technical skills to apply the science. Additionally, employees are be expected to have solid professional skills, including business and economics, ethics, and project management. Within this mode, individuals can work across a range of problems, but excel with greater mastery in selected areas, both in a domain science and technical/professional skill space.

OUTCOMES AND ONGOING CHALLENGES

Even with a continued emphasis on lifelong learning, cyclicity in employment remains a reality in the geosciences as it is in every employment sector. With the softening of the resource industries' job market in the mid-2010s, it became evident that individuals with stronger quantitative skills and additional technical skills, such as programming and database management, had greater employment resilience than geoscientists focused on a specific industry (Keane and Wilson, 2018). Two potential explanations for this trend have been considered. First, the resilient employees, by the nature of their more rigorous preparation and continued commitment to learning, potentially represent better-qualified or more motivated workers. Second, with a diverse and evolving skill set, workers are more readily able to switch career tracks when the demand for their geoscience expertise is limited. In some sectors, such as mining, where exploration and development activity can change extremely rapidly, having the ability to quickly reassign geoscience expertise can be a strategic advantage to the company and a critical approach for workers to ride through the natural business cycle pressures on employment.

Recommendations:

- Provide students with a framework for mapping future educational plans and understanding that undergraduate experience is a foundation of lifelong learning
- Make students aware of external certifications required for some geoscience employment and the availability of Continuing Education programs
- Prepare students for the changing workforce with new careers and jobs that require use of new technologies, strong quantitative and computational skills, data analytics and machine learning, interdisciplinary problem solving, and teamwork, etc.

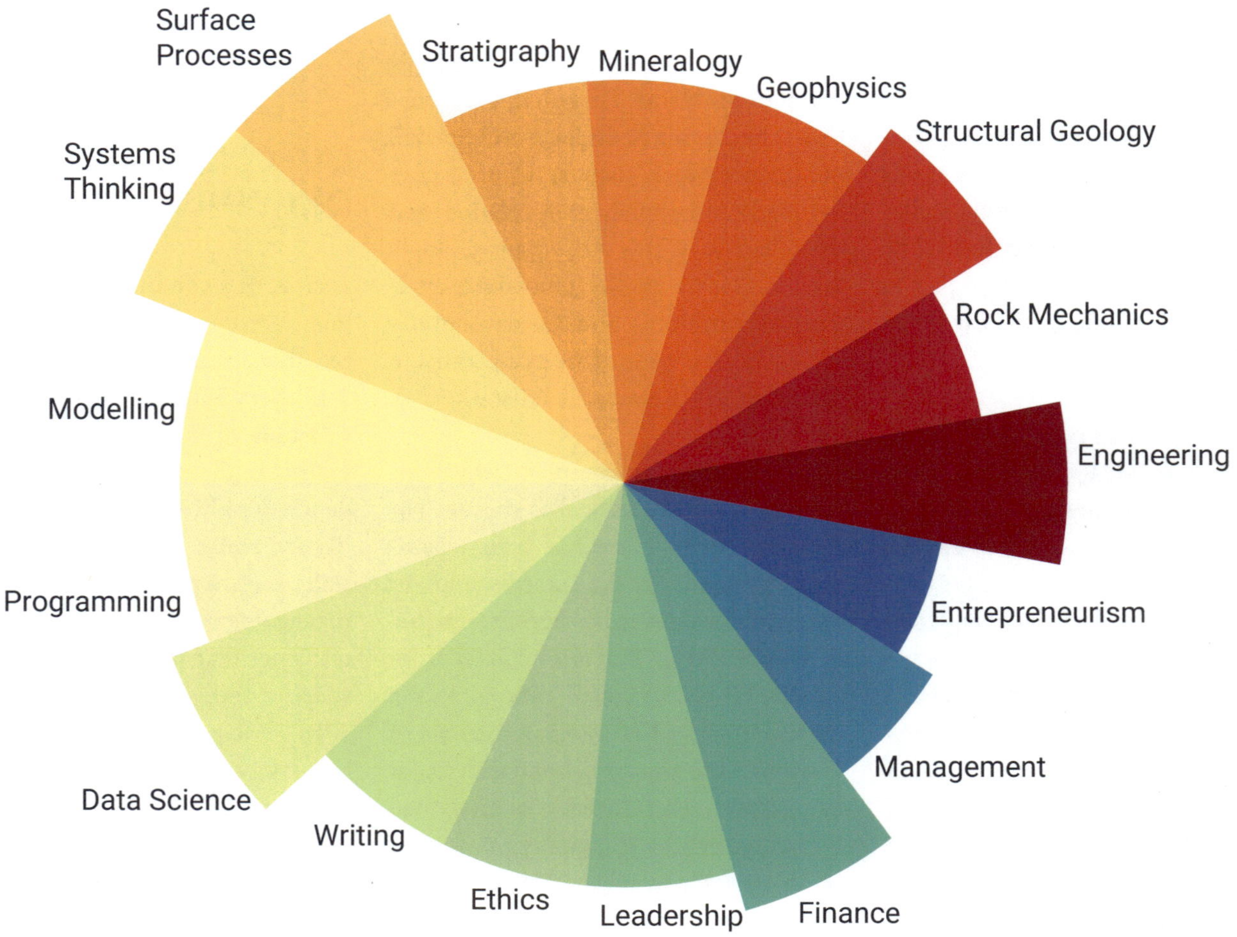

Figure 10-1: Conceptual Spectrum of Competencies for a Future Geoscientist

11. What Graduate Geoscience Education Should Accomplish

Skills and competencies needed by graduate students for successful careers should be integrated into Earth, Ocean, and Atmospheric Sciences graduate programs

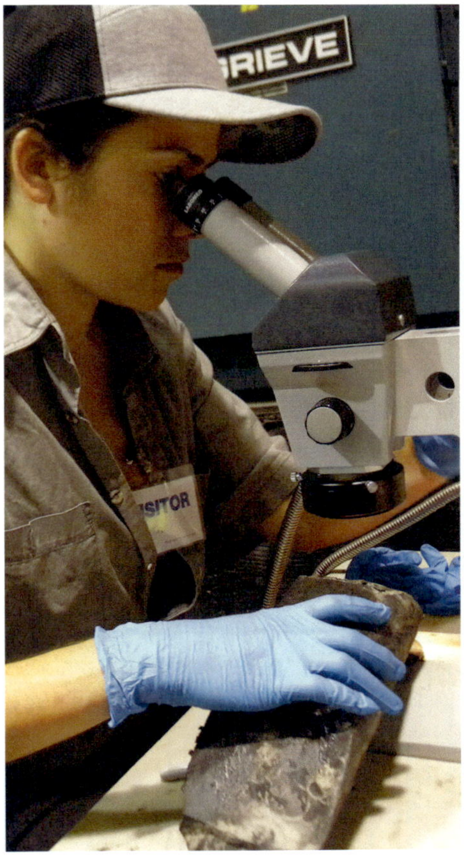

Courtesy of the Jackson School of Geosciences, University of Texas at Austin

Most geoscience employment requires a graduate education, primarily a master's degree, and for research-related careers, a doctoral degree. This continuing education builds on the undergraduate educational foundation, while adding a strong research component.

A separate NSF sponsored initiative on the "Universal Skills for Geoscience Graduate Student Success in the Workforce" investigated skills and competencies that should be part of graduate geoscience education for doctorate and master's students in Earth, ocean and atmospheric sciences. Recognizing that only about half of all STEM and geoscience doctoral students end up in academia (and under 5% of geoscience master's students; Wilson, 2018), the project's goal is to work with heads, chairs and graduate program directors on best practices and strategies for integrating employer recommended skills and competencies development into graduate programs nationally.

PROCESS SUMMARY

In the fall of 2018, a *Geoscience Employers Workshop* was held with 52 participants representing a broad spectrum of employers of geoscience graduate degree recipients in Earth, ocean and atmospheric sciences. The employers included industry, non-profits, and other organizations in the areas of weather and climate, energy and natural resources, oceans and fisheries, environment, and reinsurance and hazards. It also included government agencies including NOAA and multiple branches of NASA, as well as research labs and universities. Following the same format as the undergraduate effort, employers discussed and provided feedback to academia on the skills and competencies needed by doctoral and master's graduates for the current and future workforce. Despite the wide breadth of occupations, there was overall agreement on what graduates need and what current graduates generally lack upon entry to the workforce.

In May 2019, a Summit was held for department heads, chairs, and graduate program directors to discuss the employers' input, and strategies for supporting the development of these key skills and competencies in their graduate programs. The Summit comprised 74 participants, 60 from doctoral and 5 from master's only granting universities and colleges, plus four industry and four professional society representatives. All three branches of the geosciences (Earth, ocean, and atmospheric sciences) were represented. The heads and chairs were in general agreement with employers in terms of what graduating geoscience graduate students have and lack in terms of the identified skills and competencies. Most of the discussions focused on the best ways to address these deficiencies, and 60 individual action plans were developed and submitted. As with the undergraduate Summit effort, progress reports have be requested after 12 to 18 months.

SKILLS AND COMPETENCIES

Geoscience employers built on results of the *2015 Undergraduate Geoscience Employers Workshop* and on studies of graduate STEM education by the National Academies in 2018[41] and the Council of Graduate Schools in 2017[42]. The employers focused on the additional skills and competencies and different level of skills development that Earth, ocean and atmospheric science graduate students need for the current and future workforce. Although a large portion of the employers focused on research-oriented careers, those for whom primary research was not a focus in their sector voiced similar views.

Depth of Expertise in Core Areas

Depth of expertise in their core disciplinary areas is critical in doctoral and master's graduates, leading to good judgement and professional confidence. Employers stressed the need for good foundational skills, strong grounding in the geosciences, and course background in their chosen field, even if students switched fields after their undergraduate degrees. Developed core technical skills in their relevant areas of expertise was deemed absolutely necessary. Graduates need a deep understanding of the fundamentals and mechanics of the techniques and methods they are using. The employers also thought that in addition to breadth in their core area, graduates should have grounding across all sciences. Overall, the employers agreed that current geoscience graduates are excellent scientists with very strong technical and research skills, including laboratory and field skills, and with a solid base of knowledge of their field of geosciences.

Problem Solving and Critical Thinking

The most important skills identified regardless of discipline are problem solving and critical thinking. Graduates should be able to think logically, be flexible and open-minded, and be pragmatic in their critical thinking. The expectation for finishing graduate students, particularly doctoral recipients, was that they be independent critical thinkers who can identify and define a problem and develop and implement appropriate solutions with solid analytical and technical skills. In the workplace environment, there is usually neither the time nor need to find the complete solution, but it is critical to know whether the chosen solution is sufficient. Being able to define a sufficient solution to a problem, versus a precise and complete solution, was seen as very important.

Another aspect that employers said graduates found difficult was understanding the broader impact of their research and what decisions would be made using their findings. For any given problem, they need to understand and articulate the importance of their outcomes. Employers noted that many graduates struggle with defining a problem but could solve a problem given to them. In either case, they had difficulty identifying how to apply the solution.

Teamwork and Collaboration

Teamwork and collaboration permeate the work environment, where diversity of thought is important and valuable. The ability to work effectively with other scientists and professionals towards a project goal is critical. Beyond technical ability, being able to foster cooperation and manage conflict was crucial in teamwork. Doing so requires developing self-awareness and knowing one's own strengths and skills as well as recognizing the capabilities of the people around you. Being personally versatile matters — graduates need to be able to lead, to follow, to accept coaching and to take directions. Employers agreed that teamwork and collaboration was an area where new graduates generally lacked experience and would benefit greatly from improvement.

Leadership

Leadership involves being able to effectively guide others to accomplish goals and/or objectives in a coherent and cohesive manner. Leadership abilities are as essential in science and education as they are in business, public policy and politics. Regardless of the graduates' career directions, their ability to lead teams, groups and/or organizations is an important skill, which employers generally found lacking in their new geoscience employees. Leadership requires recognizing purpose, making commitments, identifying and

41 https://www.nap.edu/catalog/25038/graduate-stem-education-for-the-21st-century

42 https://cgsnet.org/ckfinder/userfiles/files/CGS_ProfDev_STEMGrads16_web.pdf

implementing a strategy and having the determination to accomplish one's stated goals. Leaders need to have passion and be good communicators. They also need to accept responsibility for both positive and negative outcomes, be good at listening to what others think, and appreciate the viewpoints and the accomplishments of others. Good leaders are honest and have integrity, humility, and a good set of values and principles.

Communication

Communication is a common limiting factor to success, regardless of the profession. The ability to express technical work effectively in writing and verbally to diverse audiences is critical. Audiences may include those within one's specialty, other science and engineering fields, as well as non-technical audiences such as management, the public, public policy makers, politicians and the press. Being sensitive to one's audience (i.e., reading the room), is important in finding the appropriate level to communicate effectively, and in recognizing when your audience is engaged and understands what you have to say. Complex material needs to be conveyed in a simple way, without relying on technical jargon or acronyms, and ideas expressed logically. Being able to communicate effectively about societal and/or financial impacts is important. In today's global world, effectively communicating with people whose first language is not English is also critical.

In writing, graduates need to be able to evaluate and recognize credible sources, and to have sufficient editing skills to both critically evaluate written work, and to accept and use criticism of their own writing. Employers also stressed listening skills as a critical competency: to be able to pay attention to what others say, and answer questions sensibly and logically.

Computational and Quantitative Skills

Computational and quantitative skills are essential for all types of geoscience employment. Employers thought that students entering graduate school should have the basics of statistics and higher-level math as undergraduates, including calculus, differential equations, linear algebra and statistics for communicating certainty. If not, they would need to acquire them in graduate school.

Employers noted the need for more computational skills, but within the ability to make observations. These skills include basic programming in scripted languages and being able to code and translate older code to newer and more effective systems.

Being able to develop, analyze and evaluate computational models is also a needed skill. In discussing current and future workforce needs, the ability to analyze algorithms will be important with the increase in machine learning. Keeping up with the transition to cloud computing and its unique approaches to massive scalability and parallelization will make cloud-based data manipulation and management important. For doctoral and some master's students, the employers were looking for those that embraced technology not only as users but also as creators with a willingness to engage in genuine innovation.

Data Management and Data Analytics

Of all the research skills, the one emerging skill of critical importance across the entire employer spectrum was data management and data analytics. Finishing graduate students need an awareness of data analytics, its applications, and of the processes for using data. They need to be able to work with complex and multiple

Courtesy of the Jackson School of Geosciences, University of Texas at Austin

large datasets, examining them to draw conclusions about the information they contain, including mining data to answer questions that are yet to be framed.

Data acquisition and collection requires working with different types of data and different data sources, establishing data credibility, and using available tools to access and interpret the data. Understanding and evaluating data quality and being able to appropriately use data of different qualities is essential. Data analysis and management necessitates using data effectively and being proficient at examining and synthesizing data from different perspectives (e.g., air, ground, etc.), using various types of data, and knowing the tools for the analysis and organization of data. Data analytics is an expanding field, and learning and developing new ways to manage, analyze and synthesize data will be needed.

Data integration and the merging of different types of data and information is required to solve complex geoscience problems. With a growing influx of new observations, data assimilation and sequential updating of model forecasts is necessary. Also crucial are the ability to model and know the limits of modeling, and to create visualizations and/or simulations to display and explore data. Employers also discussed the advent of machine learning, artificial intelligence (AI), robotics, immersive virtual reality data exploration, and other data science that is currently becoming significantly more important and will continue to be in the future. They also mentioned understanding how to monetize data or data valuation as a useful skill. Looking ten years to the future, employers saw all fields becoming more data centric, with the use of different programming languages, changing algorithms, and increased emphasis on tools for visualization and simulations.

Systems Thinking

Systems thinking is critical. Employers stressed the need to consider entire systems and recognize that any parts in isolation may act differently than when within the system. Thus, it is important to start at the system level and evaluate the interactions and limitations of its different parts. In solving problems, employers were looking for those who could look at the big picture first, then drill down to details, and bring that information back up to the system level. Employers agreed that systems thinking was required for all types of systems and that the Earth needs to be treated as a complex, non-linear, interactive, coupled system with interaction, linkages and feedback between different processes and parts of the system.

Project and Program Management, Business Skills

Managing projects and programs is a critical skill for success regardless of career, but one geoscience employers reported lacking among new employees. These skills include understanding budgets and project financials, managing people, time, resources and overseeing multidisciplinary projects. The specific competencies run the gamut from knowing how to run a meeting, including developing an agenda, time management, etc., to knowing the factors that drive the decision-making process.

Although proficiency was not expected, employers thought improvement in business skills was needed. Exposure to the basics of business and business operations is important, including economic, data-driven decision-making, risk analysis, uncertainty quantification, and time-value concepts, i.e., that time is money and money today is worth more than the same amount of money in the future. Innovation and entrepreneurship also play major roles in business success.

Scientific Process, Ethics and Professionalism

Employers were satisfied that their new doctoral and master's geoscience employees understood and were experienced with the scientific process of observing, characterizing, understanding, modeling, predicting, and verifying. In addition, geoscience graduates should have a good grasp of uncertainty and the scalability of space and time. Employers also discussed the need for awareness of risk and impact, the importance of simulation, and application-driven questions. Geoscience graduates should understand that research should have a clear societal purpose and be able to demonstrate that societal connection. They should have a clear understanding of research integrity as essential to science and the research process, including understanding plagiarism, self-plagiarism, proper attribution to true sources, and the ground rules for scientific citation and research. Core values are also critical, such as being trustworthy, honest, and ethical, etc.

Professional Growth

Being a life-long learner and knowing how to learn is critical, as is a diverse and adaptable skillset. New graduates should know how to search for information electronically and otherwise. Employers are looking for those with an internal drive to do well, including overcoming fear of failure and the inherent risk aversion in adopting new technology to address major problems.

Social Dynamics

Employers indicated that a general lack of social skills among most new graduates is a barrier to future success. People skills related to interpersonal and cultural behavior are very important. Personal opinions about an individual are irrelevant to professional conduct and cooperation. It is critical to be able to work with people who come from different cultures, experiences, and backgrounds.

Corporate skills were also deemed very important; academia, industry, government and business all involve different work cultures. In many environments, employees need to learn to take direction and do directed work, as well as know when innovation is encouraged and appropriate. A required corporate skill is being able to distill everything you have done and make it accessible and relevant to a CEO, manager, program director, client, or the public.

Professional Development

Students need a roadmap for professional development that starts with learning about career options and how to gain the necessary skills and knowledge to obtain their career goals. Customized roadmaps can be developed by students as they progress through their graduate career using Individual Development Plans (IDPs) (See for example, AAAS Science Careers: my IDP[43]; Section 7).

Students also need training on how to secure employment, including where to search for job openings, how to apply, and the factors that can get them hired. They need to understand the information included in a resume, a cover letter or an application depends on the type of employment being sought. Students can learn interviewing skills, how to prepare for online, phone or in person interviews, typical questions that may be asked, how they can self-market, and how they can present their unique expertise. Students also need to know how to network, including what to do and not do, and where to be or not be seen. Students should always be ready to give an "elevator speech" — a brief statement of what they have accomplished in their research and why it is significant. Today a critical piece of one's professional existence is their virtual presence or brand. How one appears on social media can affect their ability to get and keep a job. Finishing students commonly only think about their first job, but many employers said that when they interview, they are looking for people with the ability to move up and transition within an organization. Thus, it is worth investigating what the options are for advancement prior to an interview and asking about it during the interview.

Level of Competency

Both doctoral and master's students are expected to attain *mastery* of nearly all of the skills and competencies discussed above. Doctorate recipients are expected to be *experts* in terms of depth in core areas and critical thinking and problem solving. In a few areas, such as systems thinking and project management, master's recipients are only expected to be *proficient*. The level of accomplishment needed for specific skills varied to some extent with employment type. Employers across the geoscience spectrum expect competency in written and oral communication, a capacity for learning, adaptable skillsets, systems approaches, programming, simulation, data skills, problems solving and critical thinking. Employers saw their role as providing specialized job training as needed, either in house or through professional programs.

Courtesy of the Jackson School of Geosciences, University of Texas at Austin

43 https://myidp.sciencecareers.org/

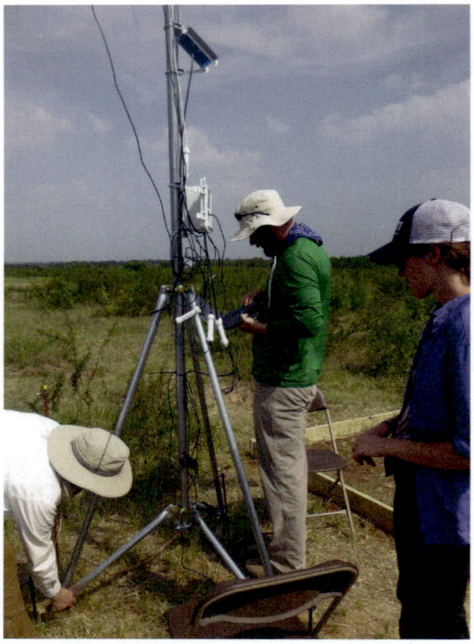

Courtesy of the Jackson School of Geosciences, University of Texas at Austin

GRADUATE EDUCATION AND FUTURE CAREERS

Most graduate education, particularly at the doctoral level, is too narrowly focused on academic research and preparing students for future academic careers. A strong doctoral researcher is distinguished by a deep technical dive into one subject, the ability to discover, own and solve a problem independently, a high level of creativity and innovation, and the ability to create new knowledge. These competencies are essential, but students also need to develop professional and personal skills valued by *both* academic and non-academic employers, including teamwork, project management, leadership, and communication. Incorporating these non-core research skills into the graduate program culture was viewed as very important, but not at the expense of losing its rich research focus. Both research and applications have transitioned to multi-, inter- and trans-disciplinary in nature. More geoscience research and occupations are critical to society, and demographics are changing to be more diverse and global. These changes require a spectrum of skills and competencies now, and the ability to continue to learn and master transferable skills.

At the 2019 Summit, heads, chairs and graduate program directors discussed ways to integrate these competencies and skills into graduate courses, research, and co-curricular programs. They recognized that faculty needed to work with their graduate students to help them identify their career aspirations, and what skills they have or need to acquire to achieve their goals. Moving toward the widespread use and support of graduate student IDP's was strongly recommended. With mentoring and acceptance of their graduate students' career goals, the needed competencies and skills can be developed without sacrificing the depth of preparation geoscience graduate students gain in research.

Recommendations:

- Graduate programs should integrate development of employer recommended skills and competencies into graduate programs while maintaining a research focus
- Faculty should support the career goals of their graduate students, recognizing that the recommended skills and competencies are essential to students' future success in both academic and non-academic careers
- Undergraduate programs should build the educational foundation needed for students who will pursue graduate degrees

12. Fostering Change in the Academic Community: Case Studies

Undergraduate program revision efforts are multi-year processes that require patience, persistence, and leadership to maintain engagement and sustain momentum.

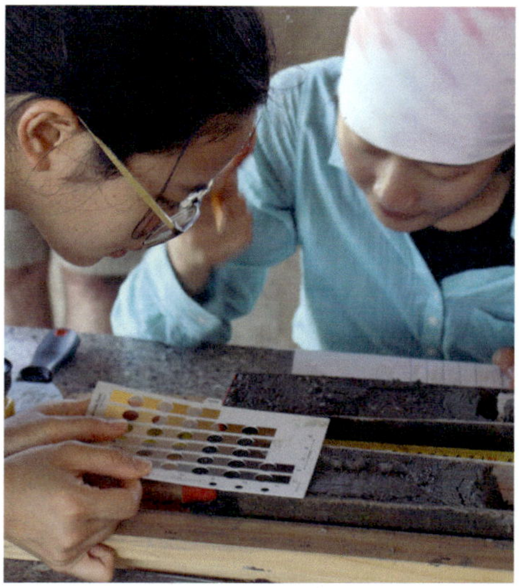

Courtesy of the Jackson School of Geosciences, University of Texas at Austin

IMPLEMENTING CHANGE

The experience of institutions that have already carried out changes, coupled with the considerable body of literature available to help businesses implement change, should prove useful to departments and programs that are to initiating curricular and pedagogical reform. Ninety-one institutions, including two-year colleges, bachelor's, master's, and doctorate-granting institutions, submitted action plans for their individual departments as part of this study; 56 of these institutions submitted progress reports on their efforts to initiate and implement change after 16 months to three years; twelve submitted a second report approximately two years after the first one. These reports provide examples of successful strategies implemented by department heads/chairs, while also describing barriers to change encountered and approaches used to overcome these challenges. They are presented and analyzed in detail in Appendix C and are described below in the context of broader strategies for successful change management.

HEADS/CHAIRS' EXPERIENCES WITH ENGAGING FACULTY IN CURRICULAR AND PEDAGOGY CHANGES

Discussions at *Summit* events and the experiences of participating heads/chairs (as discussed in progress reports on their action plans; Appendix C) brought to light successful strategies and a range of barriers, problems, and challenges to implementing changes in undergraduate geoscience programs. A common theme among participating heads/chairs in their post-*Summit* progress reports was the need for patience in starting and maintaining curricular change efforts. Building bottom-up buy-in for making change and identifying several go-to faculty to drive review and revision efforts were seen as essential, as was maintaining full transparency and gaining the assent, if not the active engagement of all faculty, as they moved forward. The chair's role in the progress of reform is largely facilitation through supporting faculty who expend time on these efforts through changes in assignment (i.e., modifying teaching loads), the allotment of course TA's, and, if appropriate, direct financial support. Heads/chairs are also positioned to align departmental curricular efforts with institution-level priorities for teaching, learning, and student success.

Successful Strategies

Many strategies were used successfully across the range of institutions. The most successful practices for evaluating and redesigning curriculums were the use of backwards design or a matrix mapping approach and faculty retreats focused on the curriculum. Heads/chairs that encouraged, supported, and incentivized professional development for faculty, including attending workshops (e.g., SAGE 2YC, Earth Educators Rendezvous) saw the greatest increase in active and experiential learning. Peer mentoring, sharing of ideas among faculty, and introduction of new ideas at retreats also helped. From the progress reports, it is clear it takes at least three years for significant changes to be made to undergraduate programs, and even then, they may still not be embedded in the departmental culture.

Other best practices developed by departments include:

- Student involvement: Showed students how courses connect, developed a curriculum roadmap that outlined expected skills and student learning outcomes, and/or developed e-portfolio programs for student assessment. Increased mentoring of diverse students and implemented training for graduate students to improve TA support of new active learning courses.
- Flipped classrooms: One department used peer mentoring by having faculty experienced in active-learning strategies co-teach with other faculty for "flipped" classes, with the other faculty taking over after one to two semesters. They started with introductory courses and ended up "flipping" three courses for majors.
- Employer involvement: Eight departments were successful in increasing internships by interacting with employers and/or brought employers in to discuss careers with students.

Departments also reported successful strategies for recruitment and retention of diverse students, easing 2YC to 4YC transfers, and program assessment. To recruit more students, some departments instituted new courses or changed introductory classes to active learning, revised marketing materials, advertised their programs, combined efforts with other STEM departments, and worked with college admissions offices on recruitment. For two- to four-year college transitions, departments eased the transfer process by increasing interactions between students and faculty at local community colleges and at four-year colleges/universities, including joint fieldtrips, student panels and social events, and development of clear transfer pathways.

Perhaps the most difficult endeavor was assessment: most departments found quantitative assessment of their changes difficult. Departments tried having faculty assess WOVN (Writing, Oral, Verbal, Numerical) in every course, exit interviews for graduates, and use of e-portfolios for gauging student achievement. One department used an alternative assessment, pre- and post-tests in all courses, bridging the previous and next course in

Courtesy of the Jackson School of Geosciences, University of Texas at Austin

the sequence. In general, departments tied their assessment of curricular changes to required annual program reviews or institutional assessments. Accreditor-motivated institutional efforts at educational assessment are both a driver and an obstacle to change, especially when these efforts are managed in a directed, top-down way. Several heads/chairs pointed out the department's advantage to using these.

Upon analysis (taking into consideration the specificity of Action Plans), having fewer geoscience faculty in a department meant more progression toward implementing their plan (See graph in Appendix C). Although faculty sizes between 5 and 10 appear to be most successful, too few faculty was a challenge as they were spread too thin to pursue a deep dive into curriculum redesign and engage in new teaching methods. With large faculty sizes (>20), it was more successful having a few faculty members take the lead with a bottom-up approach or involving early-career faculty while receiving feedback from all faculty members. For all sizes, having a faculty champion, and the support of the head/chair, was critical.

When 'Carnegie classification' is considered, 2YC, bachelor's, and master's institutions made more progress than R1 and R2 doctoral research institutions, even when R1s and R2s had equivalently small faculty sizes. Four R1s at public universities with more than 20 faculty were highly successful (see Box 12.1). All four redesigned their curriculum using a matrix approach and mapping knowledge, skills, and learning objectives across courses. Only one of the four substantially addressed pedagogy. Two provided dedicated time for a faculty member to work on program changes. All of them agreed that these changes required time and patience.

OBSERVATIONS FROM THE COMMUNITY:

Faculty have been asked to assess WOVN—Writing, Oral, Verbal, Numerical—in every course. As Chair, I have added these questions to annual performance reviews: What do you do? What have you tried? How do you know that changes you make are better? Are you still teaching what students can look up? The last question will be added to course and performance reviews in coordination with better implementation of introducing active learning methods. *(Bachelor's-granting private university)*

We also used some of our own experience, and those we learned from an NAGT site visit (the Building Strong Geoscience Departments program) to develop and implement a good course and curriculum assessment plan. This turned out to be important, as we (Geology) became a model for other departments to develop university-wide assessment plans. This year we had an accreditation site visit; this turned out well, and the vice-provost in charge of managing the assessment program, and developing the materials for our accreditation, cited these models as being critical to our successful accreditation review. This positive feedback has been affirming to our faculty and will prove (I hope) useful in resource decisions in the future. *(Master's-granting public university)*

Box 12.1: Experiences of 4 Successful Doctoral Granting R1 Public University Departments with >20 Faculty

The issues and solutions to making change are well illustrated by how these four large R1 departments were successful in making undergraduate program changes.

1. In 2017, this department had made no progress to "reconfigure our curriculum around a matrix of courses versus knowledge, skills, and ways of mind", though they had discussed it at length and it would be central to their education-based retreat in fall. In a second progress report in 2019, they say the "all-day retreat was REALLY worthwhile. We spent some time explaining the matrix approach and getting us all on the same page with regards to the Likert scaling. Then we broke into groups, which were fluid so people could join-and-leave multiple groups and real-time feedback could help us all get/stay on the same page." By 2019, they had fully developed matrices for all three of their undergraduate degree programs. They also noticed that much of the transferable skills and some important concepts were only in specific classes, so the students were only exposed once. They are now trying to get these transferable skills into majors' classes so students have a chance to develop skills throughout the curriculum. So, what made the difference? The main roadblock was time, and having a retreat was critical to their success.

2. In 2017, this department completed the "Mogk matrix" and their curriculum committee was mapping it onto their university's learning outcomes and communicating to the Dean and Provost how these were linked. They also were communicating with corporations to develop a framework to help them "translate" transcripts so the corporations knew where "teamwork" or "communication" and other skills were being developed. By 2019, they had identified gaps and ways to fill them, but for the math skills, the university curriculum course and curriculum committee rejected their plan for a Data Analytics course (despite dean approval) so they required an existing one taught in Biology. Younger faculty were pushing for more curricular reform, not just identifying and correcting gaps but thinking "out of the box." The faculty had not looked for overlap, a potential "hot button" issue, but some faculty are now passing syllabi back and forth between prerequisites and subsequent courses. The chairs' advice is: time (months, years) and patience, patience, patience. One step at a time worked, so it took us MONTHS to get through. You cannot rush this, faculty must see the results, ponder, then come back. All faculty think this was very good for the Department. They might not have "enjoyed" the process, but they realized it was worthwhile in bringing about change.

3. This department made great strides in sustaining and implementing further active learning and in "flipping" classrooms by using peer mentoring (see comments on flipping classrooms; Section 6). They also successfully implemented a new curriculum (see comments in Section 5). On their success, the chair reported, "Too many cooks can ruin the soup." "There was some initial chaos when these ideas were vetted to the entire faculty. Running all of this through a committee using a matrix 'straw-man' allowed us to get beyond the minutia of faculty concerns that can grind these efforts to a halt. One-on-one meetings also allowed us to get information quickly and in an environment that was not threatening to the faculty. Some buy-in and urgency from most faculty resulted from pressure to increase enrollment. Connecting personally and individually with professors built support and created momentum to continue the process."

Continued on page 93

Continued from page 92

4. In 2015, after a faculty member attended the *2014 Summit*, the department undertook revision of a well-regarded, but traditional, curriculum with the goal of keeping what makes the curriculum strong while enhancing its depth through improved scaffolding upon lower-division courses, increasing its flexibility, and drawing more broadly on the expertise present within the department. After considerable deliberation, a plan was proposed to the faculty in the fall of 2015. Both faculty and student feedback were solicited, and a revised plan proposed in winter of 2016 was approved, but with some reservations. The plan defined core knowledge and student learning objectives first, then integrated them into courses throughout the curriculum. They are:

   - Understand how planetary-scale processes have shaped Earth systems and habitable environments
   - Apply geologic principles to acquire data and solve problems
   - Evaluate the Earth as a set of dynamic and interacting systems
   - Become an effective geoscientist
   - Employ scientific methods via geologically informed inquiry, observation, discovery, hypothesis, testing, reason, and critical thinking
   - Synthesize and communicate knowledge of geological concepts through effective written, oral, and graphical presentation and visualization, in both collaborative and individual settings
   - Apply the tenets of professional, ethical, and responsible conduct as geoscientists

The process was derailed by major changes in leadership across the entire university spectrum including a failed search, etc. In 2017, at the Earth Educators Rendezvous, the next chair, who led the previous curriculum committee, submitted an action plan and then a progress report in 2019. The chair reported he had successfully guided debate, built consensus, and got faculty approval to implement most of the proposed changes. They did not develop tracks (e.g., geophysics or geochemistry) but might in the future, and they did not allow the capstone field course to be replaced by a senior thesis, but would consider that on a case-by-case basis. Buy-in is not 100%, more like 85%. They retained much of the present curriculum structure through the revision, and the names of the courses listed look like a relatively traditional curriculum. However, more significant changes occur within courses as they scaffold in skills, knowledge, and tools (i.e., the student learning objectives) throughout the curriculum.

The curriculum revision was under review by the University at the time of the chair's report. For the actual submission process another faculty member was granted a teaching release so they could give it their full attention. They taught all the new courses that year as a test case before the curriculum was approved. The chair's advice is: "It's a slow process. Have a committee of earlier-career faculty from diverse fields build a plan. We did start with learning outcomes, knowledge, and skills, and these did help the process. Altogether, it took us four+ years. Could have been a little faster if it weren't for other distractions affecting the department, but three years would have been the fastest it could have proceeded from start to finish."

OBSERVATIONS FROM THE COMMUNITY:

We abandoned the "Mogk" matrix when it became clear that it was going to bog us down in an overly detailed process at the expense of the big-picture reforms our curriculum needed. This decision to approach the reform in a phased way, with big-picture reform first, and course-by-course matrix analysis later, is what saved our process. *(Doctoral-granting R1 public university)*

Roadblocks and Solutions

Many challenges were similar across all types of institutions. Regardless of faculty size, many departments contended with faculty who were unwilling to change traditional curriculum or adopt new pedagogy, but with persistence many succeeded. Resistant faculty, personality issues, and intradepartmental politics interfered with attempts to make changes to undergraduate programs in about 17% of the reporting departments. In all but a few cases, these issues were resolved over time. Another 10% had to work at getting faculty buy-in and generally overcame this by either having the faculty complete the concepts and skills matrix exercise, holding faculty retreats focused on curriculum, or as a result of support and buy-in from junior or new faculty. A number of participating heads/chairs noted that an initial sense of faculty buy-in to the need to pursue curricular change and enthusiasm for major curricular reform was often followed by a loss of interest in the process when tasks such as reviewing the existing program and/or making changes to courses arose.

The *2014–2015 survey* results revealed an overall interest in making undergraduate curricular changes focusing on competencies, skills, and understanding of concepts, with 75% of the respondents indicating that their department was interested, but with only 59% indicating it was likely to happen, already in progress, or done. The *2014–2015 survey* specifically asked about obstacles and barriers to implementing research-validated pedagogies and uses of technology, and the data was analyzed by faculty and/or an administrator position (Appendix A). The *2014–2015 survey* results show that, regardless of rank or position, lack of time and/or support for developing and piloting new instructional approaches was the most important barrier to pedagogical reforms, with financial resources, instructional space design, and teaching infrastructure as additional challenges. Faculty without the rank of full professor were also concerned about the potential implications for annual performance reviews and tenure and promotion evaluations, ranging from 65% to 52% (decreasing with increased rank), indicating this was important to very important. Only assistant professors (45%) expressed concerns about student evaluations.

Box 12.2: Major Roadblocks

In some departments, faculty have different views on what constitutes a geoscience degree, which makes evaluation and redesign of curriculum difficult. For example:

A fundamental split exists among our faculty on what the future of geosciences should be. Some cannot see that geosciences need to be any different than what geology has been for decades. They want nothing added, nothing removed, nothing changed. They will concede adding something non-traditional (such a courses on climate or water!) as electives, but never at the expense of something they regard as fundamental. Other faculty feel that geosciences should change to meet current and future needs, and are trying to push for a curriculum that better connects geosciences with sustainability." The chair convened a small curriculum committee to write a short 'Learning Goals and Outcomes' document: a list of concepts, skills, and competencies that they want the undergraduate curriculum to develop. This was disseminated among the other faculty for comment and discussed at a faculty meeting; there was very little comment or discussion from faculty not on the curriculum committee. It is being incorporated into the first level courses but there has been resistance and inaction in propagating this upward through the curriculum. The chair's concern is "I've realized that we have no mechanism in the department to enforce anything. There is no way to force putting these learning goals into any course, and no mechanisms to enforce assessment of these. The biggest problem to achieving our action plan is faculty inaction. This endeavor is largely not a priority for a majority of our faculty." (Doctoral-granting R1 public university)

Another common problem was that many faculty focused only on their own courses and not the overall curriculum. It is necessary to show how intertwined courses in the curriculum are and why it is important to consider the larger picture. Even for those faculty that were receptive to programmatic changes, actualizing change was much more difficult.

Common roadblocks identified by *Summit* participants to getting buy-in from faculty for curricular or pedagogical changes and expressed in the comments in the *2014–2015 survey* and the departmental heads/chairs action plan progress reports, include:

- Resistance to inclusion of new content areas and/or the de-emphasis of other content areas affecting the faculty member's own course or the overall curriculum
- Resistance on the part of some faculty to the principles of active-learning pedagogies
- Convincing certain faculty they might need to change their courses to include important competencies/skills
- Concerns about their loss of control over course instruction and content, how and what they teach
- Concerns about increases in workload, the time and effort needed to deliver the revised program.

Many heads and chairs have advice on how they overcame these issues (see Box 12.3; Appendix C). "Patience, patience, patience", taking time, and going slow, are the most common pieces of advice. "Be persistent, encouraging, and don't expect things to change overnight." "Making slow and steady changes incrementally was received best by most faculty." "Have a vision and share that regularly with the faculty, but don't expect miracles." Showing faculty the results of the *Summits* helped spur action because it provided an externally vetted set of criteria for an undergraduate program.

Getting faculty to use a matrix to analyze the concepts and skills in their undergraduate program makes them realize their students are not learning what they think they are. The next harder step is for them to change what they are doing, collectively and individually. Advice includes: leverage institutional processes; add teaching — the use of active learning or reforming course content — as part of their annual performance review; and bring in external facilitators such as the NAGT Traveling Workshop 'Building Strong Geoscience Departments program'. Collegiality, support of leadership, departmental retreats, and the use of SERC-endorsed resources contributed to successes.

OBSERVATIONS FROM THE COMMUNITY:

We have made good progress, but slower than hoped for and not following the exact plan as initially set forth. The faculty committee agreed on a complete set of recommendations outlining a very thoroughgoing reform of our curriculum, which is now part of our strategic plan. A matrix documenting our current program was compiled, but the faculty committee was uncertain of how to use it as a tool for further planning and get faculty to make needed changes to courses. We have made little progress since then. *(Doctoral-granting R1 public university)*

OBSERVATIONS FROM THE COMMUNITY:

There is general agreement on the 'Big Picture' of where we should be heading in terms of incorporating a competency-based approach and developing problem solving and other skills into our program. The challenge is change at the course level where syllabi, lab exercises, etc., are more firmly entrenched. *(Bachelor's-granting public university)*

OBSERVATIONS FROM THE COMMUNITY:

"One of the biggest roadblocks we have as a department is inadequate facilities. We occupy an old building with limited scope for engaging in many of the active learning approaches that are increasingly popular. Lack of resources prevent us from doing anything substantial about it, so we have to work within these constraints." *(Doctoral-granting R1 public university)*

Box 12.3: Sample Advice from Heads and Chairs from Institutions Spanning 2YCs, Bachelor's, Master's and Doctorate-granting Institutions (see also Appendix C):

"Take it slow and spend the time to get faculty buy-in. Before you start, figure out how to overcome entrenched ideas regarding what constitutes a 'real' B.S. degree in geology; traveling workshops may help with this. Be prepared to do a lot of background research and bring that to the table before you engage your faculty in discussions involving major changes. Incentivize things that you can as chair." (Doctoral-granting public university)

"Start early with a rational group of faculty to gather opinions. Do not let a few individuals dominate the conversation in a general faculty meeting. Some may take a position that they think is supported, but in reality the junior faculty are afraid to confront those individuals. Set guidelines for behavior in faculty meetings." (Doctoral-granting R2 public university)

"Running all of this through a committee using a matrix 'straw-man' allowed us to get beyond the minutia of faculty concerns that can grind these efforts to a halt. One-on-one meetings also allowed us to get information quickly and in an environment that wasn't threatening to the faculty." (Doctoral-granting R1 public university)

"Use the concepts and skills matrix to your advantage, as an instrument that was nationally vetted by geoscience faculty and employers." (Master's-granting public university)

"Identifying a core group faculty to define goals and implement strategies is important. (Not all faculty members want to be involved)." (Doctoral-granting public university)

"Make sure that there are some mechanisms in place for driving and enforcing your proposed changes." (Doctoral-granting public university)

"If you are at an institution where merit pay increases are possible, you can encourage change by increasing the weight of various faculty activities in your annual merit review process. If you make 'uses-engaged learning practices' worth 20–30% of someone's teaching evaluation score in your merit review, and if you wield that evaluation sincerely and critically, you will get people to start using engaged learning practices."(Doctoral-granting R2 public university)

"Be the change. You have to model the change for others to see what works. And don't be afraid to ask what others are doing in their classrooms. Encourage them to make student-friendly decisions." (2-year community college)

"Connecting personally, and individually, with professors builds support and creates momentum to continue the process." (Doctoral-granting R1 public university)

External pressures create challenges and barriers to departments making real change and include limited campus budgets, budget cuts, hiring freezes, and low or declining enrollments. Some heads/chairs used declining enrollments to revamp introductory classes, and/or the entire curriculum, to increase class enrollments and majors. In some departments, the slow rate of progress was attributed to bureaucracy, such as the need for higher-level curriculum approval, inadequate resources and instructional space design, teaching infrastructure, and funding for new faculty, lab space, and equipment, etc. Losing one or more faculty members with no ability to replace them, and having to fill the curricular gaps left behind, exacerbated some cases.

The *2014–2015 survey* data showed fewer concerns about constraints from non-departmental sources, though 35% indicated such constraints existed in their institution.

Other external pressures causing challenges and barriers to change include:

- Upper administration imposing its own requirements and demands on departments that are at odds with their undergraduate degree program needs
- Changes in administrators, leading to a lack of clarity on institutional priorities, can slow or stop change efforts
- Inability to get buy-in from other departments on changes to cognate courses, or to offer geoscience-specific courses in cognate areas (such as geo-computation or geo-communication)
- Upper administration and university curriculum committees not understanding the need for rigorous requirements

Interestingly, while participating heads/chairs detailed their progress with faculty, only a few informed deans and other higher-level administrators about progress in curricular revision activities, or even their necessity. The few exceptions that did found it very valuable.

Some heads/chairs also talked about the importance of becoming involved in the college or university administration: "Collaborate and be involved positively with your college." "A good relationship with the administration is useful: be positive and serve on committees." A couple others solved their upper administration problems by becoming administrators.

As heads/chairs are the department's primary liaison to their university's higher administration, it is equally important for them to communicate upward about curricular review and revision efforts and the nationwide geoscience community effort to implement a consensus vision for undergraduate geoscience education. They need to show the alignment of these efforts with university-level, student learning outcomes or institutional assessment program objectives, as well as the national effort to gain approval.

OVERCOMING BARRIERS TO INSTRUCTIONAL REFORMS

Although use of research-validated teaching methods is becoming more prevalent in college geoscience courses, and instructors are placing greater emphasis on the development of key skills and competencies (Egger et al., 2019), there are nonetheless significant barriers to widespread instructional reform. University faculty members across STEM disciplines report that limited training

OBSERVATIONS FROM THE COMMUNITY:

"Work with the deans and try to get more resources. We successfully gained a new tenure-track FTE on geoscience education through a special opportunity hire. We may also get additional TA support for the new freshman-level courses we are developing. Deans and upper administration were happy with our willingness and efforts; how can they not support?" *(Doctoral-granting R1 public university)*

"Communicate curricular decisions to the Dean and develop a budget to support implementation. Deans who are not geoscientists may not be understanding of why the faculty is proposing changes, so on-going communication with the Dean about decisions being made is important." *(Doctoral-granting public university)* and lack of instructional and peer support hinder their efforts to reform their courses (Dancy & Henderson, 2010). Time is usually the most significant barrier: more than two-thirds of the *2016 National Geoscience Faculty Survey* respondents cited time constraints as the most common reason for not introducing course changes (Egger et al., 2019). The results of the *Summit* efforts support this finding. Additionally, the literature documenting the successful implementation of active strategies is unfamiliar to many geoscience instructors, and at times this literature is written in ways that are alien to those who teach geoscience courses.

No relationship apparently exists between the use of active learning strategies in geoscience courses and any institutional or demographic characteristics among geoscience programs. A national program of classroom observations documented no measurable differences in the use of active learning strategies and the type of institution (research/doctoral, master's, baccalaureate, associate), the academic level of the course (introductory vs. majors), the size of the class, or the gender of the instructor (Teasdale et al., 2017).

A significant difference was evident in the extent of active learning practices used by instructors and their cumulative professional development experiences. Instructors who participated in multiday, professional development workshops were more likely to teach using research-vetted instructional strategies than those who had not participated in such programs, or who only used related online resources, but had not attended the workshop (Teasdale et al., 2017). Those instructors who completed more than 24 hours of combined professional development programs, or who participated in programs that focused on materials topically aligned with their course content showed greater degrees of instructional reform (Viskupic et al., 2019). Nearly three-quarters (72%) of geoscience instructors interviewed noted they had changed their teaching practices as a result of their workshop participation, and were using related online resources (Manduca et al., 2017). Consequently, for interested faculty or departments, the available disciplinary professional development programs offered a means to chart an effective path to instructional reform. Heads/chairs (as reported in action plan progress reports) that encouraged and/or supported faculty attending professional development workshops saw the most successful adoption of these pedagogical reforms.

Shutterstock/Gorodenkoff

Box 12.4: Impact of COVID-19

Impact of COVID-19

The SARS-CoV-2 pandemic that began in the Spring of 2020 represents the first societally systemic disruption since at least World War 2. Though the long-term impacts on society and higher education will only be realized over the balance of the decade of the 2020s, even at this early part of the full pandemic's trajectory, it is clear that there will be lasting impacts. How will these impacts affect the future of undergraduate geoscience education and the framework presented in this document?

To address this question we will review the impacts the geoscience enterprise in the United States, which has been measured and chronicled by AGI's National Science Foundation-funded project Impacts of COVID-19 on the Geoscience Enterprise: How Permanent Will Academic Program and Workforce Changes Be? (Award #2029570).

What Happened?

As the threat from the pandemic in the U.S. became evident in March 2020, colleges and universities in the United States rapidly shifted the balance of their Spring semester to online-only teaching. As of February 2020, 91% of geoscience faculty reported teaching in-person only geoscience courses. By the end of April, no geoscience faculty reported teaching any courses only in-person, with 92% reporting teaching online-only courses. This rapid shift in teaching mode was disruptive, but it also set in motion several responses.

After the shift to online teaching, geoscience faculty reported continuing over 82% of all laboratory courses while changing the approaches to conducting those courses. Over 65% of geoscience faculty report implementing virtual laboratory processes and 55% report utilizing at-home capable activities. Additionally, 25% of geoscience faculty report adding additional computational-based activities as part of their lab instruction.

Similarly, field experiences also saw a shift in strategies, with only 5% of geoscience faculty cancelling field instruction. Rather, 74% reported utilizing virtual field trips and 40% using local field locations to effectively stay within a more controlled environment. Only 4% of geoscience faculty report continuing with normal field instruction.

Similarly, research activities evolved rapidly in the Spring of 2020, especially with respect to student research. Only 9% of geoscience students report active research being terminated, but about 20% reported planned research, such as REUs, being cancelled. For student research, the modes of activity changed, with 40% reporting shifting to virtual or computational approaches, and 22% focusing on literature review until traditional activities could resume.

The most striking aspect of the geoscience program response to the pandemic disruption was the immense agility and creativity of geoscience faculty in ensuring educational continuity of their programs. With at least 80% of labs, field experiences, and research experiences continuing in some productive form, and given the dramatic regional impacts of SARS-CoV-2 during the Spring, this represents substantial resilience by the geoscience faculty.

Emerging Sustained Changes

Heading into the Fall 2020 semester, over 50% of geoscience faculty reporting having less than one month's notice about the instructional mode for the Fall, yet several new trends began to emerge early in the semester. First, both faculty and students were acutely concerned about health and safety on campus, with well over 90% of respondents indicating it was a very large COVID-19 related concern. Interestingly, faculty and students also shared the same second and

Continued on page 100

Continued from page 99

third greatest concerns — the ability to retain academic rigor in the new teaching environments and the availability of employment for students.

All faculty and students reported that their campuses have implemented COVID-19 related restrictions to campus or classes, with 72% of geoscience courses being taught online-only. This has led to the use of new techniques such as virtual labs and field trips or shifting to computational-focused labs being more tightly integrated into the curricula for faculty.

Of note, is that where allowed, laboratory sections that can be in-person are being offered as such, with 49% of faculty indicating they are running in-person labs for Fall 2020. Likewise, for field instruction, the use of virtual field trips has dropped to 42% of faculty while 52% indicating they are conducting local field instruction for Fall 2020. In both cases, only 1% of faculty report cancelling laboratory sections and 3% cancelling field instruction for Fall 2020.

Though there is a slow shift back to traditional methods, which universally are being acknowledged as superior to alternative activities such as virtual field trips or virtual labs, the exposure to those methodologies appear to have lasting impact. A consistent message in faculty comments is that things such as virtual labs and field trips are ineffective substitutes for actual labs and field instruction, however, they represent strong, new pedagogical approaches to be used within the course as a whole, such as part of a lecture or in the introduction to lab projects.

Longer-Term Impacts

The SARS-CoV-2 pandemic has forced all faculty to evaluate their courses, consider explicitly what their expected learning outcomes are, and to evaluate new, creative approaches to ensure the continued quality of their teaching. These factors are arguably the cornerstone to the methodologies being suggested by the outcomes of the Vision and Change initiative. Will this experience make change easier? Perhaps not, but it will no longer be the first time faculty will have had to consider their courses. Additionally, as with most corporations and government agencies, universities, departments, and the faculty will be expected to be better prepared to pivot during future disruptions and thus drive for more focus on the learning outcomes of students rather than just the course scope.

The impact on higher education in general is less clear, as institutions grapple with increased costs from COVID-19 mitigation efforts, downward pressure on enrollments and tuition pricing, a more difficult fund-raising environment, and new questions on the valuation of the physical plant of the institutions. With the coinciding of an already expected drop in new enrollments in college and a shift to more online courses, COVID-19 is bringing to the forefront the debate over the value proposition of any given university program. With an expected decadal-scale period for economy recovery, graduate employability will become an even more dominant metric for value.

Employment in the geosciences has remained resilient relative to the direct impacts of COVID-19, though the energy sector had already greatly contracted in its hiring of geoscientists. According to the COVID-19 study, the geosciences experienced only a 4% annualized job loss rate during the first six months of the pandemic, with a next annual job loss rate of 1.8%, which is both far below the "full employment" standard of 5% and the prior geoscience unemployment rate of 2.1%. This resilience and the overall stability of the aggregate level of employment of geoscientists indicate that the skills are transferrable and thus a stable basis for building a career. The U.S. Bureau of Labor statistics is still forecasting aggregate growth in demand for geoscientists by 2029. So, the ability for geoscience programs to look at ways to improve the employability of their graduates will drive many programs ability to attract students and define its value proposition to university leadership. Unequivocally, higher education is being forced through a dramatic realignment, but in change exists great opportunities for those that act swiftly and strategically.

RELATIONSHIP TO SUCCESSFUL CHANGE MANAGEMENT STRATEGIES

Although the ability to change is a foundation of robust organizations, sometimes strong motivation is required to overcome institutional resistance or inertia. While this document provides guidance for vision and change in geoscience education, the geosciences community has access to the considerable literature available on implementing change in institutional settings. After studying the efforts of multiple companies to remake themselves, Kotter (2012) concludes that the most general lessons learned from successful organizational change are that the process occurs through a series of phases and generally requires a considerable amount of time. Furthermore, leaving out any of those phases frequently leads to failure in the process. Here, we relate the change framework of Kotter (2012) to the experiences reported by geoscience departments and make recommendations on how to cement that change.

Kotter's framework consists of 8 sequential steps. They are:

1. Establish a sense of urgency
2. Create the guiding coalition
3. Develop a vision and strategy
4. Communicate the change vision
5. Empower employees for broad-based action
6. Generate short term wins
7. Consolidate gains and produce more change
8. Anchor new approaches in the culture

In establishing this framework, Kotter points out that successful implementation depends on the realization that change is a multi-step process and that a strong, persevering leader is needed to steer the organization through these steps. In geoscience departments, it is the departmental head/chair who must lead the faculty through these steps from beginning to end.

Results from the *Summit* and associated workshops show that department heads/chairs have already had considerable success establishing a level of urgency around implementing curricular and pedagogical reform. Leaders need to work hard at eliminating complacency among faculty and other stakeholders.

Summit participants also reported that having a group of faculty members who strongly advocate for the change, i.e., a guiding coalition, has been key to

OBSERVATIONS FROM THE COMMUNITY:

"Because there is pressure from the university to assess and increase majors/course enrollment (e.g., falling student credit hours that have led to a lack of resources from the College), there has been some buy-in and urgency from most faculty to participate in these efforts." *(Doctoral-granting R1 public university)*

"Recognizing that classic or traditional freshmen geology classes are becoming less attractive, we were motivated to be proactive in what we teach and had the willingness to do our best to offer modern geology courses in our curriculum." *(Doctoral-granting R1 public university)*

"Our faculty, at least a group of them, were relatively willing to participate in these activities, once they understood they will strengthen our department's standing, ultimately benefit the students, and fulfill our long-term vision." *(Doctoral-granting R1 public university)*

"I think the best piece of advice is to be patient, but insistent that changes can improve our offerings and be beneficial to our students and to our program. In times of budget problems, these kinds of changes can be program savers." *(Doctoral-granting public university)*

OBSERVATIONS FROM THE COMMUNITY:

"Get a core group of faculty that can serve as representatives to various parts of the department; take your time to think it all through and do it right; consider logistics right from the outset. The core group needs passion for the project, but has to be willing to compromise in the end if necessary for agreement." *(Doctoral-granting R1 public university)*

"Empower a faculty champion. We now have a very unlikely 'hero' leading the efforts on developing the concept-skill metric and revamping our freshman-level classes. I said "unlikely' because that faculty had always put research above everything else. *(Doctoral-granting R1 public university)*

"Faculty resistance to change — but two new faculty members (out of 6) and their energy and commitment to this process are really helping us move forward." *(Master's-granting public university)*

"We've had the most success implementing change when a small group, or even just one faculty member, is willing to take the lead (rather than top-down)." *(Doctoral-granting R2 public university)*

"Identifying a core group faculty to define goals and implement strategies is important." *(Doctoral-granting public university)*

"Keeping it driven by a few faculty (bottom-up instead of top-down) I think was key." *(Doctoral-granting R1 research university)*

advancing change. Per Kotter (2012), make sure the coalition is sufficiently powerful and respected so skeptics cannot derail progress. Though engaging the skeptics is important so they feel their voices are heard.

This document presents a community vision and strategies for implementing change. Individual organizations will need to refine that vision and develop their own strategies for achieving success within their own institutional framework. Leaders using this document must help their staff fully understand the direction of change and motivate them to act, even if it is not in their short-term interest. A powerful vision is a powerful motivator.

Communicating the vision successfully comes from keeping it simple and using analogy or metaphor, repetition in multiple forums, and lots of give and take among stakeholders. This is consistent with department heads/chairs admonitions "to be patient". If the vision is under-communicated, then it will not take. It has been observed that leaders, in organizations with solid communication, are not afraid to discuss how actions and behaviors can help achieve the vision or undermine it.

The admonition to "empower employees" is about preventing perceived barriers from stopping change and is central to how department heads/chairs can remove barriers. For example, with respect to the "not enough time" complaint — how can department heads create time?

Showing faculty that change is possible via short-term wins, such as modifying a few courses or implementing some pedagogical changes and celebrating them, makes a difference. This approach makes the bigger objectives seem surmountable. Once a few short-term wins are in place, it becomes easier to consolidate and grow the process, though participants cannot be allowed to think that short-term wins represent complete victory.

A good illustration of a short-term win not being enough is provided in Box 12.1 (example #4). A department developed a revised curricular plan in 2015 that faculty approved in 2016 (short-term win). After implementation was derailed for unforeseen issues (a loss), it was restarted in 2017, and was sent to the university for approval by 2019 (a second short-term win). Only when it is approved and becomes anchored in the culture will it be fully implemented.

Anchoring the changes into culture is a huge challenge and may take many years. Kotter and others list many examples of organizations that successfully completed all the steps up until this last one, and failed. Even with near-term successes, institutional success cannot be taken for granted. Resistance will hide in many corners, and change will depend on people getting into alignment. One of the most important lessons learned is to ensure that when leaders change, the successors will continue to exemplify the new approach. The progress reports in this study showed that in cases where one department chair completed a preliminary progress report describing curricular change efforts, and was replaced by a department chair with little knowledge or interest of those efforts, much of the initial work was lost. The good news is that when the torch was passed appropriately among both leadership and staff, changes became institutionalized.

OBSERVATIONS FROM THE COMMUNITY:

"Planning is everything, but actual implementation takes time. Planning should include personal networking and faculty buy-in internally (departmental level) and externally (Senate level). The departmental plan should link to the university's strategic plan. The plan should be based on data, references, and examples and should include SWOT analyses and link to program review (self-study)." *(Master's-granting Hispanic service public university)*

OBSERVATIONS FROM THE COMMUNITY:

"Have a vision and share that regularly with the faculty, but don't expect miracles." *(Doctoral-granting R1 public university)*

Open communication and complete transparency with all stakeholders. Sharing summaries from the Summits, along with the employer-vetted concept and skills matrix, helped greatly with faculty "buy in." *(Master's-granting public university)*

OBSERVATIONS FROM THE COMMUNITY:

"Faculty buy-in may be facilitated by progressing slowly enough to allow feedback. For example, for two weeks the department filled two walls with posters listing their learning goals. This allowed everyone an opportunity to provide feedback." *(Doctoral-granting R1 public university)*

"Let everyone (faculty, staff, admin) feel like they are part of the entire process from the first steps of the initial planning efforts. It takes longer in the beginning, but having everyone on board from the start is worth the (sizable) initial investment to (hopefully) avoid issues down the road." *(Doctoral-granting R1 public university)*

OBSERVATIONS FROM THE COMMUNITY:

"We are slowly drawing more faculty into transformed courses and these experiences appear to "trickle" up to their major courses. Other faculty have seen the effectiveness and have started implementing these in *other courses*. Showing them [it works] and making them part of it [peer teaching] breaks down these preconceived notions quickly!" *(Doctoral-granting R1 public university)*

Recommendations:

- Heads/Chairs need to encourage, facilitate, and support faculty working on changes to undergraduate programs, allocate needed resources, align curricular efforts with institution-level priorities for teaching, and keep the upper administration informed of these activities and the national effort that necessitates these changes
- Review, revision, and changes to undergraduate programs and teaching are best accomplished through bottom up efforts and identifying, depending on size, one-to-a-small group of faculty to drive the effort while maintaining full transparency with the rest of the department
- Recognize that, using Kotter's change framework, it takes time and patience to implement change in undergraduate programs.

13. Engaging Stakeholders

©Shutterstock/beboy

Many individuals and organizations have a stake in the success of undergraduate geoscience education and a responsibility to accomplish the vision for the future.

In this section we examine answers to several questions related to accomplishing the community vision for the *Future of Undergraduate Geoscience Education* based on the input received from geoscience employers and academics from across the country as part of this project. They are:

- Who are the key players in facilitating change in approaches to geoscience education?
- What roles and responsibilities do those players have in the change process?
- How could/should those players interact with one another to achieve the vision?
- How do we reinforce those interactions?

The departmental head/chair is at the nexus of assorted key players and influencers including the faculty, undergraduate program director, the college dean and university administration, and external stakeholders such as the K–12/2YC community, donors and alumni, employers, professional societies, and policy makers (Fig. 13-1). It falls to the department head/chair to lead the change though, as data from this work shows, they often feel un-empowered and surrounded by barriers and obstacles, apparently set up by the key players, which limit their ability to lead change.

The *2014–2015 survey* data show that common obstacles and barriers perceived by department Heads/Chairs are:

- Faculty resistance: lack of consensus regarding the curriculum/pedagogy; lack of reward structure
- Budget cuts from above
- Lack of support/vision from above
- Lack of time
- Lack of funding

Fortunately, initial input from *Summit* and workshop participants shows that department heads/chairs can succeed in helping key players recognize their roles and responsibilities in effecting change (see Section 12; Appendix C). A head/chair can stimulate and shape change. You can educate your faculty on the general academic and employer community consensus on the conceptual understanding, skills, and competencies needed by undergraduate geoscience majors. You can empower faculty to make curriculum/course changes using the backwards design/matrix approach. You can establish academic cultures that reward innovative teaching, including promotion and tenure decisions. The key is understanding that patience and persistence are a department head/chair's power position and provides more influence and tools than readily assumed. Department heads/chairs can leverage their position in concert with responsibilities and expectations of key players to achieve the vision and the roles other stakeholders may play.

THE FACULTY

The faculty comprises disciplinary experts who have a professional obligation to keep up with the evolution and changes in their field, be that content, pedagogy, scholarship/research, changes in processes, policies or priorities of the institution, or employer demand. Department heads/chairs can take advantage of those obligations through a "carrot and stick" approach that includes leveraging "crises" such as budget cuts, building on professional pride, team building, setting and modifying workload expectations, individual goal setting, overseeing thoughtful annual performance and post-tenure review processes, and providing access and financial resources for professional development opportunities or seed projects. For example, workloads can be changed to a reduced teaching load so a faculty member may engage in curriculum design efforts to develop and pilot new instructional approaches. Department heads/chairs can engage faculty in setting the expectation that engaging in curricular change is a high priority that will be rewarded through performance evaluation processes. A budget cut can be reframed as an opportunity to make changes to the educational program so it is more effective in educating students and attractive to majors. The department heads/chairs can fund sending faculty to professional development opportunities or bring in a facilitator/consultant, either from the outside or from the University's Center for Teaching and Learning, to work with the entire faculty to enact change. Departments can offer credits and/or funding to attend these courses or workshops.

COLLEGE DEANS AND CENTRAL ADMINISTRATION

Deans and central administrators are academic professionals familiar with higher education, but likely not in how they should serve the geosciences. Their success depends on deploying their limited financial resources, time, and professional expertise to foster the meaningful improvement of academic programs. Educating them about this national initiative, and academic and employer's collective community vision for undergraduate geoscience education, demonstrates the importance of the changes your department is making. This Vision and Change document provides an opportunity and challenge to demonstrate that geoscience departments are essential, and central, parts of each institution.

In this context, a department head/chair seeking their dean's help in effecting change can continually make the case for how investment of additional resources (however small) will bring the dean

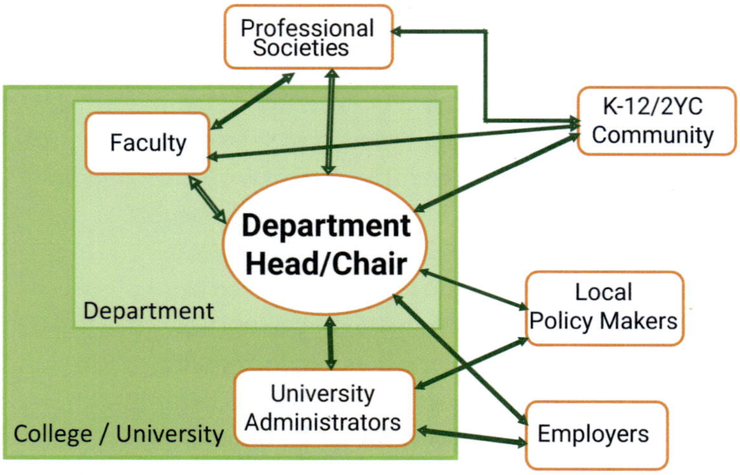

Figure 13-1: Department Chairs as Communications Nexus

reward in the form of improved metrics in the areas where they are held accountable. For example, the department head/chair has the opportunity to keep the dean continually abreast of the distinctive characteristics of the geosciences and its needs in areas such as field education, teaching laboratory space needs, and improving instructional infrastructure to better facilitate active learning.

At the same time, the department head/chair can keep the dean informed on successes that reflect on the dean, such as student employment opportunities or the department's role in community engagement, natural resources, and sustainability. This should create a reinforcing loop for the dean to "reward" the department, recognizing the reward may be intermediate to long term. Typically, if a dean is well-informed and energized about a department's contribution, this success will trickle up to the Provost's office. However, it can be useful to become involved in university committees to increase the visibility of the department and its good work.

Note that the *2014–2015 survey* data from this project yielded relatively little information on how deans and upper administration can help, rather than hinder, department heads. The paragraph above represents the insights of the writing team, some who hold, or have held, these roles.

K–12 COMMUNITY

K–12 science and Earth science teachers can use their expertise and insight on how the Next Generation Science Standards are being implemented, particularly in middle and high schools, to help college and university faculty understand what to expect from incoming students and what pre-service teachers need from their undergraduate education. Faculty can help teachers interest students in the geosciences by developing and providing examples and resources useful in middle and high school classrooms, and by giving talks or lectures. Collaboration between teachers and faculty at the different institutions can increase dual credit or On-Ramps courses and ease the transition between high school and college.

2YC COMMUNITY

Community Colleges have diverse administrative structures, therefore understanding the local structures increases the chance of a successful collaboration. Many regions may have several community colleges in a 4YC's service area, which can make communicating with 2YC stakeholders a challenge. Reaching out to determine what geoscience degree plans, technical programs, and Earth science courses they offer and what their student enrollment/population is, can help focus on the specific goal of a 4YC and help with collaboration efforts.

Community college administrators that oversee geosciences courses or programs need to ensure their faculty are aware of the efforts and trends discussed in this document and what is occurring on a national level regarding 2YC geoscience education. Geoscience 2YC faculty in turn need to educate unaware administrators. Funding for faculty to attend workshops, conferences, and meetings is vital to learning more about these efforts and making important connections. Universities are more likely to have faculty with larger networks and collaborate across multiple institutions on projects, often with the aid of travel funds written into research grants to help disseminate their information and ideas. 2YC faculty, on the other hand, are often less connected with their peers across the country and therefore miss out on the successful and replicable initiatives within the national 2YC community. The SAGE 2YC and *Summit* series of events has helped change this dynamic. The most effective method of broadening 2YC faculty connections and awareness of national strategies is to enable them to attend local, state, and national geoscience meetings and workshops.

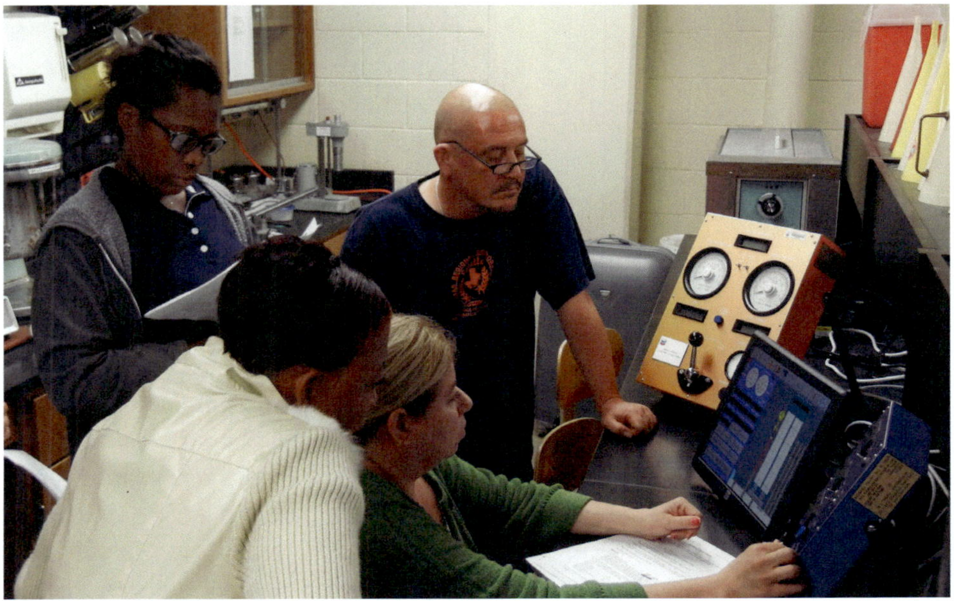

Courtesy of the Jackson School of Geosciences, University of Texas at Austin

mana5280 on Unsplash

Collaboration between faculty at 2YC and 4YC colleges and universities is critical to ease the transition for students transferring from 2YCs to 4YCs that can increase 4YC enrollment and diversity. Joint fieldtrips, faculty exchanges of seminars and lectures, research collaborations involving students, and other mechanisms can increase collaboration between faculty and students at both institutions. 2YC faculty have expertise and insights on effective teaching and working with diverse communities that is helpful for 4YC faculty.

MUSEUMS AND OTHER INFORMAL EDUCATION SETTINGS

Museums, science centers, and other types of informal learning settings play important roles in promoting the geosciences by enhancing knowledge about the Earth and the connection of geoscience processes to everyday life. They provide educational opportunities, many through active learning, and can stimulate student interest in the geosciences and reinforce concepts and scientific practices learned in traditional educational settings. Heads/chairs can encourage faculty and students to use museums and their online learning resources to extend learning opportunities and reinforce geoscience concepts.

ALUMNI/ DONORS

Understanding where undergraduate students go after graduation, where they are employed, and how they keep in touch throughout their careers is important. Engaging in a long-term ongoing discussion or longitudinal survey is needed to understand what was most and least helpful and what was missing in their undergraduate studies relative to the career they have undertaken.

Alumni and development offices may maintain this information, but departments need to posses this data to effectively engage with alumni on these programmatic topics. Facebook, LinkedIn, ResearchGate, etc., are all helpful in finding alumni, some of whom, may be interested in financially supporting specific initiatives and enjoy hearing what and how their department is doing.

EMPLOYERS

Non-academic employers have professional geoscientists who likely have a much more applied and business orientation than university faculty. In some cases, they may be alumni representing local businesses engaging with the University, or others hiring geoscience students. Their professional success depends in part on the ability to identify and hire geoscience graduates who can be efficiently on-boarded to contribute to the mission of the firm or government agency.

Participating employers noted that the department head/chair may find they can reach out to employers to help with students' education in many ways. Employers can help by providing samples, data sets, case studies, and opportunities for end-to-end learning through conceptualization, collection of data, analysis, and presentation. Company employees and retirees may be willing to help or teach classes or problem-oriented short courses.

Employers are the best resource for promoting career awareness. By fostering communication and engagement between employers and academia, students and faculty gain a better understanding of the skills and competencies needed for success. In addition to having alumni and other employers on advisory boards, invite them as speakers for talks, seminars, and/or classes to discuss 'real-world' perspectives, applications, their jobs, and career opportunities. See if they will do short or mock interviews, or give advice on resumes and interviewing skills, online or in person. Some employers will do externships exposing students to what is involved in working for that company. Another source of career information is returning interns who can talk about their experiences, the corporate culture, and values.

If you have research symposia where students present, either verbally or in poster sessions, invite employers to participate, and if appropriate, be judges. This is mutually beneficial — students get feedback from external professionals, and employers learn about the department's program and have an opportunity to see students in a non-interview setting.

PROFESSIONAL SOCIETIES

The geoscience societies, such as GSA, AGU, AAPG, SEG NAGT and AGI, play a central role in being career hubs — one that students can start engaging with as a student and continue participating with professionally well past retirement. The societies are the keepers of the profession; their members *are* the society and their values and priorities reflect those of the profession. The societies can disseminate, promote, and achieve the community vision of the *Future of Undergraduate Geoscience Education* and evolve their own activities as well. Possibilities include: spearheading follow-on meetings and workshops; providing professional development and/or mentorship for students, faculty, and other geoscientists; modifying programming of meetings and shifting editorial focus of journals; and developing or continuing of programmatic activities reflecting the evolved view of geoscience resulting from the *Summits* (Box 13.1).

OBSERVATIONS FROM THE COMMUNITY:
Our Friends and Alumni Network do mock interviews, using a "speed dating" approach, with our students when they come to campus for their board meeting. They also serve as judges for our annual student research poster symposium, providing students a chance to interact with professional geoscientists and get their input. In turn, our alumni have a chance to see what the students can do in a non-interview setting. *(Doctoral-granting R1 public university)*

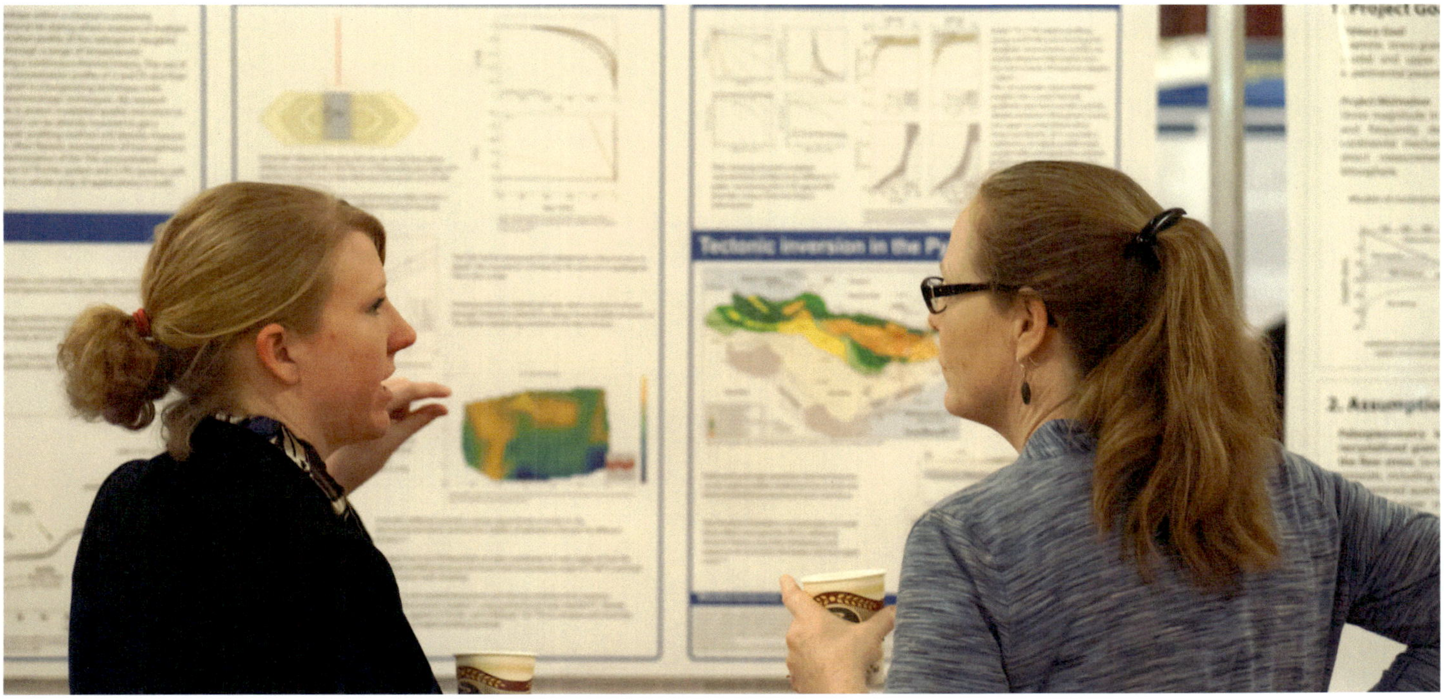

Courtesy of Jackson School of Geoscience, University of Texas at Austin

Box 13.1: Actions for Professional Societies

At the Summits and Geoscience Employer Workshop, specific suggestions were made for constructive actions by professional societies:

- Provide more experiences to connect employers with students
- National and regional meeting sessions with career related information, poster sessions with companies and employers, and networking opportunities
- Provide venues for employers to interact with students in one-on-one conversations
- Connect students to willing mentors in their disciplines by developing mentoring programs and expanding current ones — i.e., AGU's Mentoring365 virtual mentoring and GSA's Schlemon and Mann Mentoring programs. Broaden a mentoring network to include more career paths e.g., Congressional Fellows, Knauss Marine Policy Fellows could be public policy mentors
- Promote externship and internship opportunities — act as a clearinghouse
- Help develop and provide competency certification or badging, curricula, and/or other educational programs and activities
- Offer summer courses and field experiences for students and young professionals
- Offer short courses and workshops to develop professional and technical skills at national or regional meetings
- Expand opportunities for students to learn to effectively communicate to different audiences. For example, at professional meetings hold sessions designed specifically for students to present their research with verbal and/or written feedback from professionals or a general audience (public, K–12 teachers, policy fellows, etc.)
- Develop competency, certification/accreditation, and/or badging programs for desired skills
- Create opportunities for industries and other employers to engage with academia; most faculty have little contact with employers
- Increase student focus and encourage continuing membership
- Strengthen the role of their student chapters as catalysts for the various reforms proposed, especially in co-curricular opportunities
- Develop funding to send more students to professional meetings
- Provide opportunities for young professionals to be mentors and interact with students at meetings
- Focus on supporting all students and young professionals and not competing with them as members
- Design membership enrollment processes similar to engineering societies, such as batch memberships for all students in a department and having a network of student organizations across the US at major universities schools
- Identify pan-society student and departmental support opportunities and programs

The vision of the future of the geosciences is that of an integrated continuum from student to professional. The societies historically serve both populations, with a primary focus on the professional and a reactive program approach to students. This discrete approach has likely hampered their ability to retain student members through their transition into the profession, with most societies reporting fewer than 10% of student members becoming professional members on a continuing basis. However, these same people appear to rejoin in their mid-30s. Societies need to view members not as discrete categories, but rather as valued individuals transitioning through phases of their life as a geoscientist.

It would behoove the societies to start accepting the new geoscience majors as simply inexperienced geoscientists and look at their structures, activities, and how they touch a person across all phases of their career. The graduate-to-5 years' experience phase, described as "emerging geoscientist" by SEG and by other societies as the "awkward teenage years of a geoscientist", is where emerging professionals need to be supported like students and provided opportunities like professionals.

PUBLIC POLICY AND THE PUBLIC

The professional societies, as well as individuals, can engage in change relative to public policy. Historically, specific initiatives pushing discipline-specific changes through the federal system have had little effect. For example, prior efforts to build sustained funding sources for mining engineering programs faced challenges. Such focused efforts run counter to most government investment approaches. The traditional argument, even if flawed, is that if the need is critical, then the market (a.k.a. industry) will find ways to support a solution.

Geosciences also faces a problem of scale in developing human capital nationally. The geosciences tend to be a small player in the primary GDP contribution and are often allocated as an expense for mitigation or regulatory compliance. When coupled by widespread skilled worker shortages such as pilots, truckers, nurses, etc., the national and public priority will align with the more immediate and tangible challenges.

On the federal level, current rhetoric in 2020 related to higher education costs and concerns about the return on investment of a college degree needs to be followed closely by all in the academic community. If changes are enacted to the current federal aid system, this will impact all programs. These questions have been further complicated by the impact of the COVID-19 crisis and impact on college and university operating budgets. The question is how to prepare for change, and especially how to address the current phase where higher education as a business is having to prospectively align itself for its predicted future, which currently is utilitarian in nature with an increased focus on graduate economic outcomes.

Beyond the federal influence through student aid and research grants, most impacts on higher education occur at the state and local level. These policy bodies are traditionally responsive to local opinion and conditions. If the geoscience community proactively and visibly demonstrates applied solutions to local problems with measurable economic impact, then local policymakers may favor supporting geoscience programs and students. However, providing solutions, and not simply identifying compliance issues, is

©iStock/Tyler Fairbank

critical. This makes the public perceive geoscience as being current and part of the team building strong local communities. The department's core activity will then be demonstrating how these proposed reforms will build a responsive skilled workforce that addresses specific local issues such as flooding, land stability, and economic resources, especially when they have been impacted within the "political memory" of the election cycle.

It is also critical to engage with local school and state education boards since changes to K–12 curriculum and requirements can have a major impact on your incoming students and the future teachers you are educating.

Departments also need to be cognizant of the composition of their board of regents, or similar bodies. Often these are composed of local business and political leaders, whose first interest will be to understand a program's local impact versus proclamations of national and international stature and placement of graduates at distant organizations.

CATALYZING THE DIALOGUE BETWEEN ACADEMIA AND EMPLOYERS TO FOSTER FUTURE EVOLUTION

Undergraduate geoscience programs need to evolve with the science, societal issues, and the needs of employers of future graduates. Surveys by AGI have shown that graduate outcomes versus employer expectations are well-aligned, especially in the areas of applied and professional skills (Houlton, 2015). Employers and academics generally agreed on skills as well in the *2014–2015 survey* (Summa et al., 2017). However, strong dialogue and a collaborative relationship between industry and departments, is critical to enabling responsive evolution in geoscience programs. Such relationships also invest industry partners in the health and effectiveness of the geoscience programs from which they draw their talent.

During the *2015 Geoscience Employers Workshop*, the elements of what constitutes effective and viable collaboration and dialogue were discussed at length. The specifics will vary by circumstance, but some of the targeted ideas are detailed below.

FACULTY/EMPLOYER SABBATICAL SWAPS

Pair with employers for a faculty sabbatical to pursue research and experience within that environment. Likewise, enable industry partners to have resident time in the geoscience department to conduct research, teach classes, and work with students. Such arrangements might be designed synchronously as a faculty/professional swap or could be opportunistic based on specific circumstances.

ENGAGEMENT OF ALUMNI/PROFESSIONALS IN COMMITTEES AND ACTIVITIES

Engage local and other professionals to serve on review committees, career advising councils, or similar functions to make persistent connections with local employers and those that hire your students. Though there is often a tradition of tapping alums for such roles, consider engaging non-alums to build out a broader support network and to provide insight and experiences that differ to avoid a completely inward focus that can isolate a program from changes. Where feasible, participation from non-regional alumni, or employers, can also serve to broaden the support network.

PROFESSIONAL SOCIETY CHAPTERS

Some of the professional societies such as AIPG, have chapters that connect students to the local professional community. Utilize these chapters to engage the department at an organizational level with the regional or national bodies. Examples of involvement at an organizational level might include active involvement in regional conferences or similar events or building critical networking opportunities between students, faculty, and industry. Departments need to incentivize the broader faculty to engage with these groups as it will form a lasting relationship beyond the duration of student residence time.

Most professional development short courses offered by geoscience societies are conducted by industry members. These courses and lectures are often offered at the local chapter level. If direct student participation in the short course is not possible, leveraging relationships through campus chapters, etc., could be engaged to repeat a short course for students.

PROFESSIONAL LICENSING OF FACULTY

Identify key faculty to obtain a professional geologist license in your state (if available). This license opens up consulting opportunities, encourages currency in applied geoscience issues of concern to local employers, and in many states, enables students to begin to earn "professionally supervised" time toward a professional geologist license themselves, possibly even before graduating.

SEEK "LOST" ALUMNI

Many alumni disappear and do not communicate with their alma mater departments. With only 50% of working geoscientists being formal members of professional societies, there is a large body of practicing geoscientists and employers who are not in the normal communication flow of the geosciences. Making your department their anchor will help them engage and provide unique connections and relationships.

PROVIDE REAL DATA SETS AND PROBLEMS

Departments benefit from accessing real data from actual work projects or case studies. Employers at the *2015 Geoscience Employers Workshop* indicated that providing such resources should be rather simple, but the single largest barrier is that they are not directly asked. Shared datasets allow students and faculty to collaborate with the providing company on solutions to problems, including the different levels of science required for a complete answer vs. a "sufficient" answer in a budget-constrained environment where business decisions may be made without complete information.

JOINT CLASS-BASED RESEARCH OPPORTUNITIES

Work with local industry to identify projects or problems that might be appropriate for a class-based research opportunity. Local priority or undercapitalized research problems can be identified by companies and will provide students with real experiences and a value-added return for the collaborating company.

INTERNSHIPS

Internship experiences are critical for building work and professional experience by students. Even internships that lead a student to decide a particular area of work is not what they want has as much value as the one where they find their passion.

PAIRED SENIOR THESIS EXPERIENCES WITH INTERNSHIP

One of the hallmarks of many geoscience programs is the senior thesis project. Traditionally, many thesis projects are approached like a master's research project. AGI studies on library access notes measurable demand for access to senior theses by engineering and consulting firms because they often represent the only pre-existing, detailed data for many locales in the United States. Departments should approach local employers to identify projects of immediate or speculative use that could be conducted as a joint internship/senior thesis experience.

American Geosciences Institute

OBSERVATIONS FROM THE COMMUNITY:
Geoscience programs will grow and thrive when their graduates can demonstrate that their knowledge and skills are grounded in innovative thinking and strong preparation for their role in a dynamic society.

FACILITATE TECHNOLOGY APPLICATION OPPORTUNITIES

Research will often yield new and novel techniques, technologies, or methods that can be patented or licensed. Collaboration with industry partners, crafted with assistance from a university technology transfer office, could not only lead to a patentable invention, but the initial market capital to yield a revenue-generating product for the faculty inventor and the department. In fields with thin patent records like the geosciences, a ready-market definition is critical to designing appropriate business plans, and with licensing preferences, for instance, industry partners can be key.

Recommendations:

- Heads and chairs: encourage and support your faculty to complete necessary curricular review and revision efforts and in adopt new instructional approaches

- Heads and chairs: advocate to the dean for support of your department's innovations related to meeting our new community educational standards, articulating their relationship to institutional measures of student success

- Faculty engaged in instructional reform: take or make opportunities to participate in professional development experiences, the more specific to the geosciences and their courses the better

- Funding agencies, professional societies, and others: support and/or offer a generous menu of professional development experiences for faculty and students to better prepare them for their careers, for adopting reformed teaching practices, and for curricular enhancement efforts

- Academic departments and geoscience employers: seek to develop and maintain interactive professional relationships with each other, focusing ultimately on improving the abilities and accomplishments of bachelor's geoscience graduates

The community vision for the future of undergraduate geoscience education articulated in this report is a roadmap for making critical, positive changes to undergraduate programs over the next decade. Our learning environments and curricula must evolve to confront future geoscience challenges and prepare students to enjoy a vibrant and successful career. We urgently need to reconsider our role in educating the next generation of geoscientists for the health of our profession and success of our students.

References

Archer, R. S., Davis, F., Ebanks, S. C., Gragg, R. D. S., 2019, HBCUs Broadening Participation in Geosciences (A Journey Through InTeGrate), *in* Gosselin D., Egger A., Taber J. (eds) Interdisciplinary Teaching About Earth and the Environment for a Sustainable Future, AESS Interdisciplinary Environmental Studies and Sciences Series. Springer, Cham., doi: 10.1007/978-3-030-03273-9_17.

Asera, R., 2001, Calculus and Community: History of the Emerging Scholars Program, The College Board, New York, https://files.eric.ed.gov/fulltext/ED562582.pdf.

Asher, P., 2019, Career Compass, American Geosciences Institute Workforce Program, https://www.americangeosciences.org/workforce/compass.

Asher, P., Keane, C., and Cantner, K., 2018, Geoscience Workforce Pathways: A Holistic Approach at Preparing for Careers in the Geosciences. Geological Society of America Abstracts with Programs. v.50, n.6, doi: 10.1130/abs/2018AM-316760.

Ashley, M., Cooper, K. M., Cala, J. M., Brownell, S. E., 2017, Building Better Bridges into STEM: A Synthesis of 25 Years of Literature on STEM Summer Bridge Programs. CBE Life Sci Education, v.16 n.4:es3. doi:10.1187/cbe.17-05-0085.

Association of State Boards of Geology, 2016, Fundamentals of Geology Examination (FG) as an "assessment examination", ASBOG report, https://asbog.org/documents/Assessment%20Exam%20Handout--February2016.pdf.

Baber, l. D., Pifer, M. J., Colbeck, C. and Furman, T., 2010, Increasing Diversity in the Geosciences: Recruitment Programs and Student Self-Efficacy, Journal of Geoscience Education, v.58, n.1, p. 32–42, doi: 10.5408/1.3544292.

Banilower, E. R., Smith, P. S., Weiss, I. R., Malzahn, K. A., Campbell, K. M., Weis, A. M., 2013, Report of the 2012 National Survey of Science and Mathematics Education, Horizon Research, Inc., Chapel Hill, NC, 390p., http://horizon-research.com/NSSME/2012-nssme/research-products/reports/technical-report.

Banks-Santilli, L., 2015, Feet on campus, heart at home: first-generation college students struggle with divided identities, The Conversation, https://theconversation.com/feet-on-campus-heart-at-home-first-generation-college-students-struggle-with-divided-identities-42158.

Barrett, H., 2010, Electronic Portfolios in STEM, STEMTech Conference, https://www.slideshare.net/eportfolios/eportfolios-in-stem.

Barstow, D., Geary, E., and Yazijian, H., eds., 2002, Blueprint for Change: Report from the National Conference on the Revolution in Earth and Space Science Education, Snowmass, CO, June 21–24, 2001, 100p.

Beane, R., McNeal, K. S., Macdonald, R. H., 2019, Probing the National Geoscience Faculty Survey for Reported Use of Practices That Support Inclusive Learning Environments in Undergraduate Courses, Journal of Geoscience Education, v.67 n.4, p. 427–445, doi: 10.1080/10899995.2019.1621714.

Beckham, J. T., Simmons, S., and Farre, J., 2015, The Freshman Research Initiative as a Model for Addressing Shortages and Disparities in STEM Engagement, *in* Rubinstein, U. A., and Peterson, M. A., eds., Directions for Mathematics Research Experiences for Undergraduates, Chapter 10, p. 181–194.

Bernard, R. E., and Cooperdock, E. H., 2018, No progress on Diversity in 40 Years, Nature Geoscience, v.11 n.5, p. 292–295.

Bishop, J. L., and Verleger, M. A., 2013, The Flipped Classroom: a Survey of the Research, *in* ASEE National Conference Proceedings, Atlanta, GA.

Boyer, E., 1990, Scholarship Reconsidered: Priorities of the Professorate, Carnegie Foundation for the Advancement of Teaching.

CLEAN (Climate Literacy and Energy Awareness Network)., 2018, Guidance in Teaching About Climate and Energy, https://cleanet.org/clean/literacy/.

Coble, R., 2019, Pedagogy for Professional Schools and Students. Accessed May 29, 2019. https://cft.vanderbilt.edu/guides-sub-pages/pedagogy-for-professional-schools-and-students/.

Cribb, W., 2018, Documenting Student Success Through E-portfolios, AGU-AGI Heads and Chairs program video presentation, https://www.americangeosciences.org/webinars/documenting-student-success-through-e-portfolios.

Crouch, C. H. and Mazur, E., 2001, Peer Instruction: Ten Years of Experience and Results. American Journal of Physics, v.69, p. 970–977.

Czajka, C. D., and McConnell, D., 2016, Situated Instructional Coaching: a Case Study of Faculty Professional Development, International Journal of STEM Education, v.3 n.1, p. 10, doi: 10.1186/s40594-016-0044-1.

Dancy, M., and Henderson, C., 2010, Pedagogical Practices and Instructional Change of Physics Faculty, American Journal of Physics, v.78, n.10, p. 1056.

Danielson, C. and Abrutyn, L., 1997, An Introduction to Using Portfolios in the Classroom, Association for Supervision and Curriculum Development, Alexandria, VA.

Derting, T. L., and Ebert-May, D., 2010, Learner-Centered Inquiry in Undergraduate Biology: Positive Relationships with Long-Term Student Achievement, CBE-Life Sciences Education, v.9, p. 462–472.

Desimone, L. M., and Garet, M. S., 2015, Best Practices in Teachers' Professional Development in the United States, Psychology, Society & Education, v.7, n.3, p. 252–263.

Dickerson, D., Womack, V. B., Stwalle, C. S., 2014, Can an Engineering Summer Bridge Program Effectively Transition Underrepresented Minority Students Leading to Increased Student Success?, 131st ASEE Annual Conference & Exposition, Paper ID #10977.

Eddy, S. L., and Hogan, K. A., 2014, Getting Under the Hood: How and for Whom Does Increasing Course Structure Work?, CBE-Life Science Education, v.13, p. 453–468.

EERE (US Office of Energy Efficiency and Renewable Energy), 2017, Energy Literacy Framework, https://www.energy.gov/eere/education/energy-literacy-essential-principles-energy-education.

Egger, A. E., 2019, The Role of Introductory Geoscience Courses in Preparing Teachers - and all Students - for the Future: Are we Making the Grade?, GSA Today, v.29, p. 4–10.

Egger, A. E., Viskupic, K., and Iverson, E. R., 2019, Results of the National Geoscience Faculty Survey (2004–2016), National Association of Geoscience Teachers, Northfield, MN, 82p. https://serc.carleton.edu/NAGTWorkshops/CE_geo_survey/.

ESLI (Earth Science Literacy Initiative), 2010, Earth Science Literacy Principles, http://www.earthscienceliteracy.org/.

Eynon, B., and Gambino, L. M., 2018, Catalyst in Action: Case Studies of High-Impact ePortfolio Practice, Sylus Publishing, 380p.

Eynon, B., and Gambino, L. M., 2017, High-Impact ePortfolio Practice, Sylus Publishing, 256p.

Freeman, S., Eddy, S. L., McDonough, M., Smith, M. K., Okoroafor, N., Jordt, H., and Wenderoth, M. P., 2014, Active Learning Increases Student Performance in Science, Engineering, and Mathematics, Proceedings of the National Academy of Sciences, v.111, p. 8410–8415.

Freeman, S., Haak, D., and Wenderoth, M. P., 2011, Increased Course Structure Improves Performance in Introductory Biology, CBE-Life Sciences Education, v.10, p. 175–186.

Gajjar, N. B., 2013, The Role of Technology in 21st Century Education, International Journal of Research in Education, v.2, p. 23–25.

Garet, M. S., Porter, A. C., Desimone, L., Birman, B. F., and Yoon, K. S., 2001, What Makes Professional Development Effective? Results from a National Sample of Teachers, American Educational Research Journal, v.38, n.4, p. 915–945, doi: 10.3102/00028312038004915.

Gast, I., Schildkamp, K., and Veen, J. T. Van Der, 2017, Team-Based Professional Development Interventions in Higher Education: A Systematic Review, Review of Educational Research, v.87, n.4, p. 1–32, doi: 10.3102/0034654317704306.

Gates, A. E., McNeal, K., Riggs, E., Sullivan, S., and Dalbotten, D., 2019, New Developments in Diversity and Inclusiveness in Geosciences, Journal of Geoscience Education, v.67, n.4, p. 285–286, doi: 10.1080/10899995.2019.1671713.

Gilbert, L. A., Gross, D. S., and Kreutz, K. J., 2019, Developing Undergraduate Students' Systems Thinking Skills with an InTeGrate Module, Journal of Geoscience Education, v.67, n.1, p. 34–49, doi: 10.1080/10899995.2018.1529469.

Gonzales, L., 2010, Underrepresented Minorities in the U. S. Workplace, Geoscience Currents 34, https://www.americangeosciences.org/geoscience-currents/underrepresented-minorities-geoscience-occupations.

Gosselin, D. C., Egger, A. E., and Taber, J. J., eds., 2018, Interdisciplinary Teaching About Earth and the Environment for a Sustainable Future, Springer.

Graham, M. J., Frederick, J., Byars-Winston, A., Hunter, A. B., and Handelsman, J., 2013, Increasing Persistence of College Students in STEM, Science, v.341, p. 1455–1456.

Grissom, A. N., Czajka, C. D., and McConnell, D. A., 2015, Revisions of Physical Geology Labs to Increase the Level of Inquiry: Implications for Teaching and Learning, Journal of Geoscience Education, v.63, n.4, p. 285–296.

Gross, B., Marinari, M., Hoffman, M, DeSimone, K; Burke, P., 2015, Flipped @SBU: Student Satisfaction and the College Classroom, Educational Research Quarterly, v.39, n.2, p. 36–52.

Haak, D. C., Hille RisLambers, J., Pitre, El, and Freeman, S., 2011, Increased Structure and Active Learning Reduce the Achievement Gap in Introductory Biology, Science, v.332, p. 1213–1216.

Handelsman, J., Miller, S., and Pfund, C., 2007, Scientific Teaching, W. H. Freeman, New York.

Henderson, C., Beach, A., and Finkelstein, N., 2011, Facilitating Change in Undergraduate STEM Instructional Practices: An Analytic Review of the Literature, Journal of Research in Science Teaching, v.48, p. 952–984.

Henderson, C., and Dancy, M., 2007, Barriers to the Use of Research-Based Instructional Strategies: The Influence of Both Individual and Situational Characteristics, Physical Review Special Topics - Physics Education Research, 3:020102-1–14.

Hofstra, B., Kulkarni, V. V., Galvez, S. M. N., He, B., Jurafsky, D., and McFarland, D. A., 2020, The Diversity-Innovation Paradox in Science, Proceedings of the National Academy of Sciences, v.117, n.17, p. 9284–9291.

Houlton, H., and Keane, C., 2017, Can Geoinformatics Catalyze Improved Gender and Ethnic/Racial Diversity in Geoscience and Computer Science? Geological Society of America Abstracts with Programs. v.49, n.6. doi: 10.1130/abs/2017AM-300831.

Houlton, H., 2015, Geoscience Career Master's Preparation Survey Report, American Geosciences Institute.

Ireton, F., Mogk, D. W., and Manduca, C. A., eds., 1996, Shaping the Future of Undergraduate Earth Science Education: Innovation and Change Using an Earth System Approach, AGU, Washington, DC.

Johnson, J. and Reynolds, S., 2005, Concept Sketches—Using Student- and Instructor-generated, Annotated Sketches for Learning, Teaching, and Assessment in Geology Courses. Journal of Geoscience Education, v.53, n.1, http://www.nagt.org/nagt/jge/abstracts/jan05.html#v53p85.

Karsten, J. L., 2019, Insights from the OEDG Program on Broadening Participation in the Geosciences, Journal of Geoscience Education, v.67, n.4, p. 287–299, doi:10.1080/10899995.2019.1565982.

Kastens, K. A., Zrada, M. and Turrin, M., 2019, What Kinds of Questions do Students ask While Exploring Data Visualizations?, Journal of Geoscience Education, doi: 10.1080/10899995.2019.1675447.

Kastens, K., and Manduca, C., 2017, Leveraging the Power of a Community of Practice to Improve Teaching and Learning about the Earth, Change: The Magazine of Higher Learning, v.49, p. 14–22.

Kastens, K. A., Manduca, C. A., Cervato, C., Frodeman, R., Goodwin, C., Liben, L. S., Mogk, D. W., Spangler, T. C., Stillings, N. A., and Titus, S., 2009, How Geoscientists Think and Learn. Eos, v.90, n.31, p. 265–272.

Keane, C., and Wilson, C., 2018, The Mid-21st Century Geophysics Workforce: How Today's Trends Across Geoscience Impact Geophysics Human Resources of the Future, doi: 10.1190/segam2018-2992425.1.

Kotter, J. P., 2012, Leading Change, Harvard Business School Press, Boston, 208p.

Lage, M. J., Platt, G. J., and Treglia, M., 2000, Inverting the Classroom: A Gateway to Creating an Inclusive Learning Environment, Journal of Economic Education, v.31, n.1, p. 30–43.

Libarkin, J. C., Anderson, S. W., 2006, The Geoscience Concept Inventory: Application of Rasch Analysis to Concept Inventory Development in Higher Education *in* Applications of Research Measurement in Science Education, ed. Liu, X. and Boone, W., JAM Publishers, p. 45–73.

Lopatto, D., 2007, Undergraduate Research Experiences Support Science Career Decisions and Active Learning, CBE-Life Science Education, v.6, p. 297–306.

Macdonald, R. H., Manduca, C. A., Mogk, D. W., Tewksbury, B. J., 2005, Teaching Methods in Undergraduate Geoscience Courses: Results of the 2004 On the Cutting Edge Survey of US Faculty, Journal of Geoscience Education, v.53, p. 237–252.

MacDonald, H., Manduca, C., Mogk, D., and Tewksbury, B., 2004, On the Cutting Edge: Improving Learning by Enhancing Teaching in the Geosciences *in* Invention and Impact: Building Excellence in Undergraduate Science, Technology, Engineering and Mathematics (STEM) Education, AAAS, Washington, DC, p. 233–240, https://www.aaas.org/sites/default/files/08_Cre_App_McDonald.pdf.

Malchuk, D., 2018, Re-shaping the Future, 2018 Annual Meeting, Prospectors and Developers Association of Canada.

Manduca, C., Iverson, E., Tewksbury, B., Luxenburg, M., Macdonald, H., McConnell, D. A., Mogk, D., and Tewksbury, B. J., 2017, Improving Undergraduate STEM Education: The Efficacy of Discipline-Based Professional Development, Science Advances, 3:e1600193, doi: 10.1126/sciadv.1600193, https://advances.sciencemag.org/content/3/2/e1600193.

Manduca, C. A., Mogk, D. W., Tewksbury, B., Macdonald, R. H., Fox, S. P., Iverson, E. R., Kirk, K., McDaris, J., Ormand, C. and Bruckner, M., 2010, On the Cutting Edge: Teaching Help for Geoscience Faculty, Science, v.327, n.5969, p. 1095–1096.

Manduca, C. A. and Mogk, D. W., eds., 2006, Earth and Mind: How Geologists Think and Learn about the Earth, v.413, Geological Society of America.

Martinsen, O. J., Talwani, M., Levander, R., Dengo, C., Barkhouse, B., Dunn, J. F., Link, C., Mosher, S., Tatham, R., Orcutt, J., Paul, D., Talley, R., 2012, A U. S. Human Resource Challenge for Earth Science Education and Energy Exploration and Exploitation, The Leading Edge, 30th Anniversary Volume, v.31, p. 714–716, doi:10.1190/tle31060714.1.

Matthews-DeNatale, G., 2013, Are We Who We Think We Are? Eportfolios As A Tool For Curriculum Redesign, Online Learning, v.17, n.4, doi: 10.24059/olj.v17i4.395.

Mazur, E., 1997, Peer Instruction: A User's Manual, Prentice Hall, New Jersey, 253p.

McConnell, D. A., 2019, Research-Based Instructional Reform in the Geosciences: Building a Community of Practice *in* Levers for Change: An Assessment of Progress on Changing STEM Instruction, AAAS, Washington, DC, 200p.

McConnell, D. A., Chapman, L., Czajka, C. D., Jones, J. P., Ryker, K. D. and Wiggen, J., 2017, Instructional Utility and Learning Efficacy of Common Active Learning Strategies, Journal of Geoscience Education, v.65, n.4, p. 604–625.

McConnell, D., et al., 2006, Using ConcepTests to Assess and Improve Student Conceptual Understanding in Introductory Geoscience Courses, Journal of Geoscience Education, v.54, n.1, p. 61–68.

McConnell, D., et al., 2003, Assessment and Active Learning Strategies for Introductory Geology Courses, Journal of Geoscience Education, v.51, n.2, p. 205.

McFadden, R. R., Viskupic, K., and Egger, A. E., 2019, Faculty Self-Reported Use of Quantitative and Data Analysis Skills in Undergraduate Geoscience Courses, Journal of Geoscience Education, doi: 10.1080/10899995.2019.1700595.

McTighe, J. and Wiggins, G., 2012, Understanding by Design Framework, ACSD White Paper, https://acsd.org.

Miller, A. R. and Kastens, K. A., 2018, Investigating the Impacts of Targeted Professional Development Around Models and Modeling on Teachers' Instructional Practice and Student Learning, Journal of Research in Science Teaching, v.55, n.5, p. 641–663, doi: 10.1002/tea.21434.

Mogk, D., 2016, Some Reflections on Curriculum Design and Program Assessment, Presentation at the Heads and Chairs Summit on the Future of Undergraduate Geoscience Education, Jan 8–10, 2016, Austin, TX.

Mogk, D., 2015, Creating the Graduates You Want, 2015 Earth Educators Rendezvous Mini-Workshop, Boulder, CO, https://serc.carleton.edu/earth_rendezvous/2015/mini_workshops/mw10/.

Mogk, D., 2014, A Curriculum by Design, Part II: Student Learning Outcomes and Program Assessment, Earth and Mind Blog, https://serc.carleton.edu/earthandmind/posts/curriculum_desi2.html.

Mogk, D., 2013, An Earth Science Curriculum by Design, InTeGrate Geoscience and the 21st Century Workforce Essay Collection, https://serc.carleton.edu/integrate/workshops/workforce2013/essays/mogk.html.

Mosher, S., Bralower, T., Huntoon, J., Lea, P., McConnell, D., Miller, K., Ryan, J., Summa, L., Villalobos, J., and White, L., 2014, Future of Undergraduate Geoscience Education: Summary Report for Summit on Future of Undergraduate Geoscience Education, School of Geosciences Faculty and Staff Publications, p. 1127, https://scholarcommons.usf.edu/geo_facpub/1127, https://www.jsg.utexas.edu/events/files/Future_Undergrad_Geoscience_Summit_report.pdf.

Moss, E., and Cervato, C., 2016, Quantifying the Level of Inquiry in a Reformed Introductory Geology Lab Course, Journal of Geoscience Education, v.64, p. 125–137.

Murphy, T. E., Gaughan, M., Hume, R., and Moore, Jr., S. G., 2010, College Graduation Rates for Minority Students in a Selective Technical University: Will Participation in a Summer Bridge Program Contribute to Success?, Educational Evaluation and Policy Analysis, v.32, n.1, p. 70–83, doi: 10.3102/0162373709360064.

National Park Service, U. S. Department of Interior, 2009, National Park Service Comprehensive Survey of the American Public, Natural Resources Report NPS/NRSS/NRR-2011/431.

NASEM (National Academies of Sciences, Engineering, and Medicine), 2015a, Science Teachers' Learning: Enhancing Opportunities, Creating Supportive Contexts. The National Academies Press, Washington, DC, doi: 10.17226/21836, https://www.nap.edu/catalog/21836/science-teachers-learning-enhancing-opportunities-creating-supportive-contexts.

NASEM (National Academies of Sciences, Engineering, and Medicine), 2015b, Integrating Discovery-Based Research into the Undergraduate Curriculum: Report of a Convocation., The National Academies Press, Washington DC, doi: 10.17226/21851.

McFarland, J., Hussar, B., Zhang, J., Wang, X., Wang, K., Hein, S., Diliberti, M., Cataldi, E. F., Mann, F. B., Barmer, A., NECS (National Center for Education Statistics), 2019, The Condition of Education 2019, NCES 2019–144.

National Academy of Engineering, 2013, Messaging for Engineering: from Research to Action, The National Academies Press, Washington, DC, doi: 10.17226/13463.

National Academy of Engineering, 2008, Changing the Conversation: Messages for Improving Public Understanding of Engineering. The National Academies Press, Washington, DC, doi: 10.17226/12187.

NGSS Lead States, 2013a, Next Generation Science Standards: For States, By States. The National Academies Press, Washington, DC, https://www.nextgenscience.org/.

NGSS Lead States., 2013b, Next Generation Science Standards: For States, By States. Classroom Sample Tasks, The National Academies Press, Washington, DC, https://www.nextgenscience.org/classroom-sample-assessment-tasks.

NGSS Lead States, 2013c, Next Generation Science Standards: For States, By States. Evidence Statements, The National Academies Press, Washington, DC, https://www.nextgenscience.org/evidence-statements.

NGSS Lead States, 2013d, Next Generation Science Standards: For States, By States, MSS-ESS1-1 Performance Expectation, The National Academies Press, Washington, DC, https://www.nextgenscience.org/topic-arrangement/msspace-systems.

Noel-Levitz Research, 2009, Institutional Brand and Parental Influence on College Choice, 10p.

NRC (National Research Council), 2015, Guide to Implementing the Next Generation Science Standards. The National Academies Press, Washington, DC, doi: 10.17226/18802, https://www.nap.edu/catalog/18802/guide-to-implementing-the-next-generation-science-standards.

NRC (National Research Council), 2014, Developing Assessments for the Next Generation Science Standards, The National Academies Press, Washington, DC, doi: 10.17226/18409, https://www.nap.edu/catalog/18409/developing-assessments-for-the-next-generation-science-standards.

NRC (National Research Council), 2012, A Framework for K–12 Science Education: Practices, Crosscutting Concepts, and Core Ideas, The National Academies Press, Washington, DC, doi: 10.17226/13165, https://www.nap.edu/catalog/13165/a-framework-for-k-12-science-education-practices-crosscutting-concept.

NRC (National Research Council), 2000, Inquiry and the National Science Education Standards: A Guide for Teaching and Learning, The National Academies Press, Washington, DC, doi: 10.17226/9596.

OLN (Ocean Literacy Network), 2015, Ocean Literacy Framework, http://oceanliteracy.wp2.coexploration.org/.

PCAST, 2012, Engage to Excel: Producing One Million Additional College Graduates with Degrees in Science, Technology, Engineering, and Mathematics. Office of the President, Washington, DC.

Pollock, S. J. and Finkelstein, N. D., 2008, Sustaining Educational Reforms in Introductory Physics, Physical Review Special Topics - Physics Education Research, v.4, n.1, p. 010110.

Ricchezza V., and Vacher, H. L., 2017, A Twenty-Year Look at "Computational Geology," an Evolving, In-Discipline Course in Quantitative Literacy at the University of South Florida, Numeracy, 10, doi: 10.5038/1936-4660.10.1.6.

Rosaz, T., 2013, How to Build Local Capacity, EAGE GeoSkills 2013.

Russell, S. H., Hancock, M. P. & McCullough, J., 2007, Benefits of Undergraduate Research Experiences, Science, v.316, p. 548–549.

Ryker, K. D., and McConnell, D. A., 2017, Assessing Inquiry in Physical Geology Laboratory Manuals, Journal of Geoscience Education, v.65, p. 35–47, doi: 10.5408/14-036.1.

Ryan, J. G., Nocita, B., Rodriguez, E., Schackne, M., Hull, J., Vacher, H. L., Connor C. B., and Rains, M., 2017, Twenty-plus Years of a Successful Employer-Academic Partnership in Southwest Florida, GSA Annual Meeting, Seattle, WA.

Ryan, J. G. and Schackne, M., 2016, Developing Successful Academia-Industry Partnerships, AGU/AGI Heads and Chairs Webinar Program, February 13, 2016, https://www.americangeosciences.org/workforce/webinars/developing-successful-industry-and-academia-partnerships-example-florida.

SAGE 2YC, 2019, Provide Career Information, https://serc.carleton.edu/sage2yc/careers/.

Salamis, G., 2017, Artificial Intelligence, Big Data and Crowdsourcing: Mineral Exploration in the Digital Age, 2017 Annual Meeting, Prospectors and Developers Association of Canada.

Santiago, A. M., Einarson, M. K., 1998, Background Characteristic as Predictors of Academic Self-Confidence and Academic Self-Efficacy Among Graduate Science and Engineering Students, Research in Higher Education, v.39, p. 163–198, doi: 10.1023/A:1018716731516.

Savina, M. E., Buchwald, C., Edward, B., David M., and Boardman, S. J., 2001, A Skills Matrix as a Geology Department Curriculum Planning Tool, GSA Paper No. 79-0.

Sherman-Morris, K. and McNeal, K. S., 2016, Understanding Perceptions of the Geosciences Among Minority and Nonminority Undergraduate Students, Journal of Geoscience Education, v.64, p. 147–156.

Simmons, S., 2014, One Institution's Approach: How the University of Texas at Austin Merges Research and Teaching Through the Freshman Research Initiative (93.1), FASEB Journal, v.28, n.1, supplement.

Spencer, S. J., Logel, C., and Davies, P. G., 2016, Stereotype Threat, Annual Review of Psychology, v.67, p. 415–437.

St. John, K., ed., 2018, A Community Framework for Geoscience Education Research. National Association of Geoscience Teachers. Retrieved from https://doi.org/10.25885/ger_framework/15.

Stes, A., Min-Leliveld, M., Gijbels, D., and Van Petegem, P., 2010, The Impact of Instructional Development in Higher Education: The State-of-the-Art of the Research, Educational Research Review, v.5, n.1, p. 25–49, doi: 10.1016/j.edurev.2009.07.001.

Strayer, J. F., 2012, How Learning in an Inverted Classroom Influences Cooperation, Innovation and Task Orientation, Learn Environ Research, v.15, p. 171–193.

Summa, L., Keane, C., and Mosher, S., 2017, Meeting Changing Workforce Needs in Geoscience with New Thinking about Undergraduate Education. GSA Today, v.27, n.9, p. 60–61.

Sunal, D. W., Hodges, J., Sunal, C. S., Whitaker, K. W., Freeman, L. M., Edwards, L., Johnston, R. A. and Odell, M., 2001, Teaching Science in Higher Education: Faculty Professional Development and Barriers to Change, School Science and Mathematics, v.101, n.5, p. 246–257.

Tinto, B., and Engstron, C., 2009, Learning Better Together: The Impact of Learning Communities on the Persistence of Low-Income Students, Opportunity Matters, v.1, p. 5–21.

Teasdale, R., Viskupic, K., Bartley, J. K., McConnell, D. A., Manduca, C., Bruckner, M., Farthing, D., and Iverson, E., 2017, A Multidimensional Assessment of Reformed Teaching Practice in Geoscience Classrooms, Geosphere, v.13, p. 260–268.

Tewksbury, B. J., Manduca, C. A., Mogk, D. W., and Macdonald, R. H., 2013, Geoscience Education for the Anthropocene, GSA Special Paper 501, p. 189–201.

UCAR (University Corporation for Atmospheric Research), 2007, Atmospheric Science Literacy Framework, https://scied.ucar.edu/atmospheric-science-literacy-framework.

Ufnar, J. A., Kuner, S., and Shepherd, V. L., 2012, Moving Beyond GK–12, CBE-Life Sciences Education, v.11, n.3, p. 239–247, doi: 10.1187/cbe.11-12-0119.

Vacher, H. L., 2013, Teaching with Spreadsheets Across the Curriculum, SERC Pedagogical Services Collection, https://serc.carleton.edu/sp/ssac/.

Vacher, H. L., 2000, A Course in Geological Mathematical Problem Solving, Journal of Geoscience Education, v.48, p. 478–481.

Vacher, H. L., and Emily L., 2010, Spreadsheets Across the Curriculum, 1: The Idea and the Resource, Numeracy, v.3, n.2, doi: 10.5038/1936-4660.3.2.6.

Viskupic, K., Egger, A. E., McFadden, R. R., and Schmitz, M. D., 2020, Comparing Desired Workforce Skills and Reported Teaching Practices to Model Students' Experiences in Undergraduate Geoscience Programs, Journal of Geoscience Education, doi: 10.1080/10899995.2020.1779568.

Viskupic, K., Ryker, K., Teasdale, R., Manduca, C., Iverson, E., Farthing, D., Bruckner, M. Z. and McFadden, R., 2019, Classroom Observations Indicate the Positive Impacts of Discipline-Based Professional Development, Journal for STEM Education Research, p. 1–28.

Wenner, J. M., and Baer, E. M. D., 2015, The Math You Need When You Need It (TMYN): Leveling the Playing Field, Numeracy, v.8, doi: 10.5038/1936-4660.8.2.5.

Wenner, J. M., Burn, H., and Baer, E. M. D., 2011, The Math You Need, When You Need It: Online Modules that Remediate Mathematical Skills in Introductory Geoscience Courses, Journal of College Science Teaching, v.41, p. 16–24.

Wenner, J. M., and Baer, E. M. D., 2010, The Math You Need When You Need It, SERC Educational Resource Collection, https://serc.carleton.edu/mathyouneed/.

Wieman, C., Perkins, K., & Gilbert, S., 2010, Transforming Science Education at Large Research Universities: A Case Study in Progress, Change: The Magazine of Higher Learning, v.42, n.2, p. 6–14.

Wiggins, G., and McTighe, J., 2005, Understanding by Design, 2nd Edition. ASCD, Alexandria, VA, https://acsd.org.

Whitmeyer, S., Pyle, E., De Paor, D., Atchison, C., Bentley, C., Piatek, J., 2016, Digital Technologies that Enhance Teaching and Learning in Classroom and Field Environments, GSA Abstracts with Programs, v.48, n.7.

Wilson, C. E., 2019, Status of the Geoscience Workforce 2018, American Geosciences Institute.

Wilson, C. E., 2018, Status of Recent Geoscience Graduates 2017, American Geosciences Institute.

Wolfe, B. A., and Riggs, E. M., 2017, Macrosystem Analysis of Programs and Strategies to Increase Underrepresented Populations in Geosciences, Journal of Geoscience Education, v.65, n.4, p. 577–593, doi: 10.5408/17-256.1.

Wulff, A. and May, M., 2013, Intentional Learning for Career Success, The Professional Geologist, Mar/Apr 2013.

Wysession, M. E., 2014, The Next Generation Science Standards: A Potential Revolution for Geoscience Education, Earth's Future, v.2, p. 299–302, doi: 10.1002/2014EF000237.

Appendix A:
Results from the 2014 Summit and 2014–2015 Survey

FUTURE OF UNDERGRADUATE GEOSCIENCE EDUCATION SUMMIT; JANUARY 2014

Participants

About 200 educators representing a broad spectrum of the undergraduate geoscience education community:

- R1, R2, and R3 research universities with undergraduate programs, doctoral/professional universities, terminal master's programs, 4-year private and state colleges and 2-year community colleges
- Faculty, heads & chairs, education researchers
- Industry & professional society representatives
- Working in small groups with collective presentations

Outcomes

- Collective agreement on most points
- Online survey developed to assess larger community views and status of community efforts
- Many questions were designed for academics and non-academics did not see those pertaining only to departments

©Michael Collier, from the AGI ESW Image Bank

DEMOGRAPHICS OF THE 2014–2015 SURVEY

- 463 respondents
  - 68 were participants in the Summit (15%)
  - 357 Academics (77%)
  - 79 Industry (17%; 47 energy/oil/gas, 21 mining/minerals exploration, 11 environmental/hydrogeology/engineering)
  - 14 Government (3%)
  - 7 Other (1.5%)
  - 5 Professional Society Representatives (1%)
  - 315 Male (68%)
  - 148 Female (32%)

- Ethnicity
  - 91% White/Caucasian
  - 2% Hispanic
  - 1% Asian
  - 1% American Indian/Alaska Native
  - 5% Other
  - 0% Black/African American and Native Hawaiian/Other Pacific Islander

- In all reported results, all non-academics are listed as employers.

Academic Demographics

- Department Faculty size
  - 19% under 5
  - 31% 5–12
  - 21% 12–20
  - 12% 20–30
  - 17% over 30

- Institution type
  - 16% 2-year community colleges
  - 12% 4-year private
  - 30% 4-year state
  - 34% public research universities
  - 7% private research universities
  - 1% Hispanic-Serving Institution

- Position
  - 10% nontenure track/adjunct
  - 11% assistant professor
  - 23% associate professor
  - 40% professor
  - 11% department chair/head
  - 4% dean or upper administration

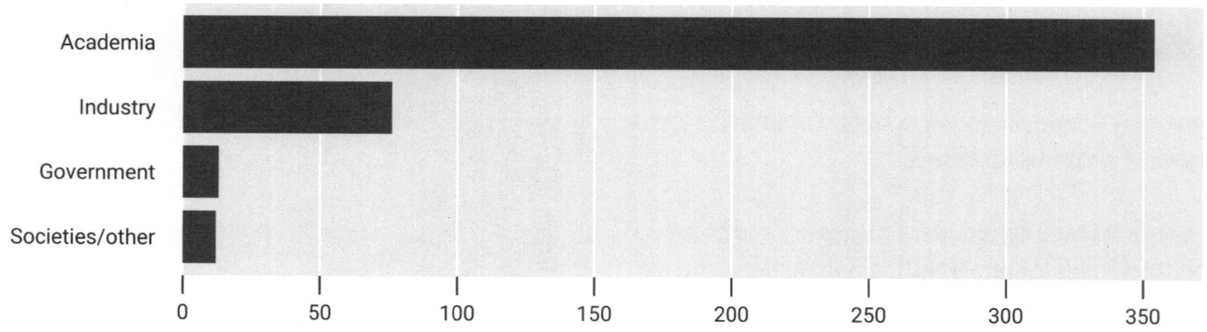

Figure A-1: Participants by Sector
Total number of survey and summit participants by sector

EDUCATION EMPHASIS

For Undergraduate Geoscience Education

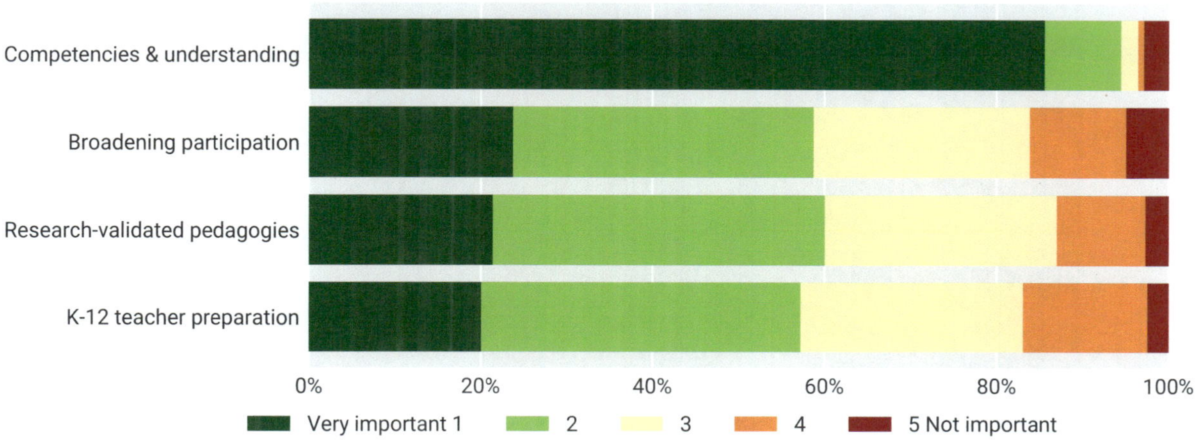

Figure A-2: Issue Importance in Undergraduate Geoscience Education

Survey Question: What are the most important issues from your perspective in terms of undergraduate education?

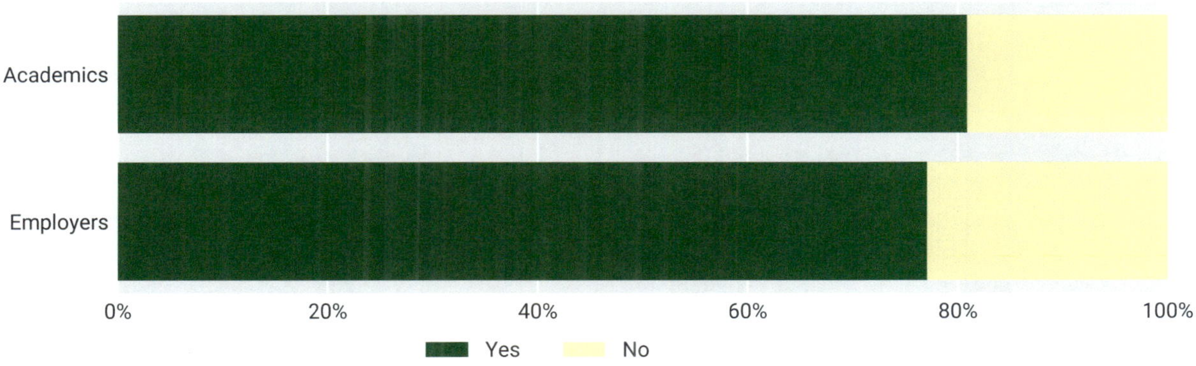

Figure A-3: Are Competencies More Important Than Specific Courses?

Survey Question: A major outcome of the summit was an agreement that developing competencies, skills, and conceptual understanding were more important than making sure specific courses were taught at an undergraduate level. Do you agree with this approach?

IMPORTANCE OF TEACHING KEY GEOSCIENCE CONCEPTS

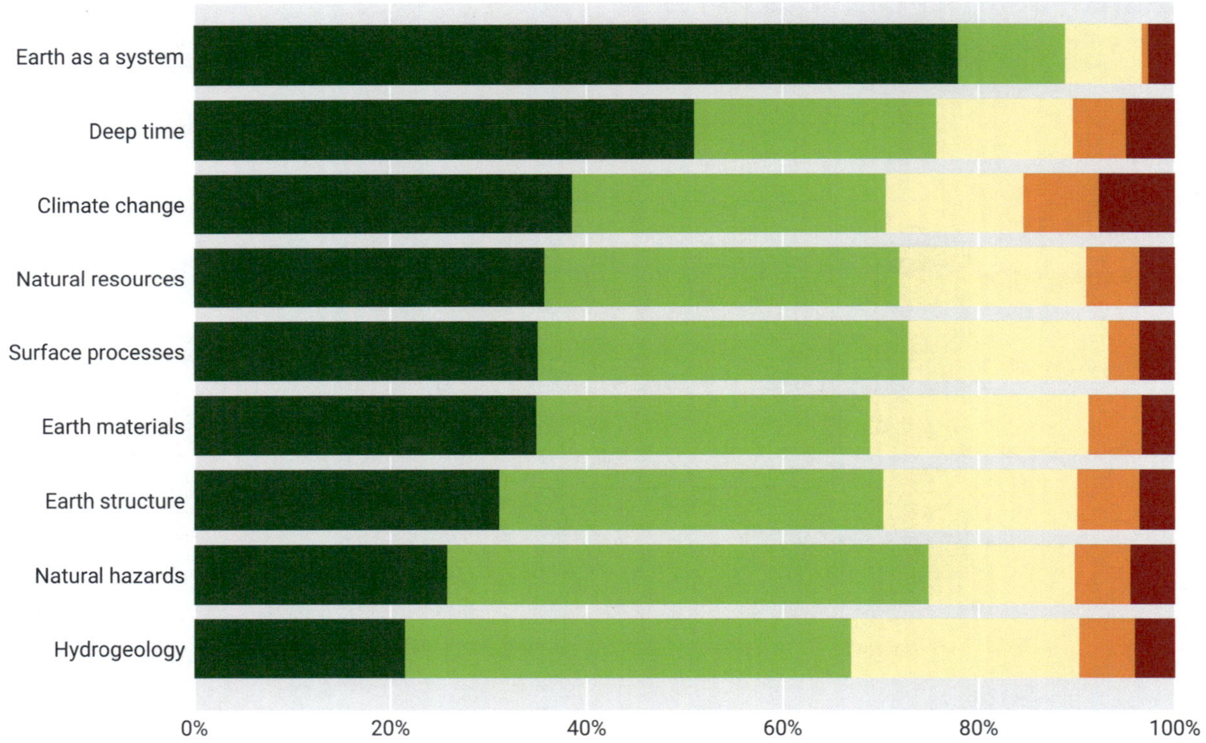

Figure A-4: Importance of Key Geoscience Concepts

Survey Question: What are the key geoscience concepts that need to be addressed in undergraduate geoscience education?

Concept Importance Breakdowns by Respondent Employment Category

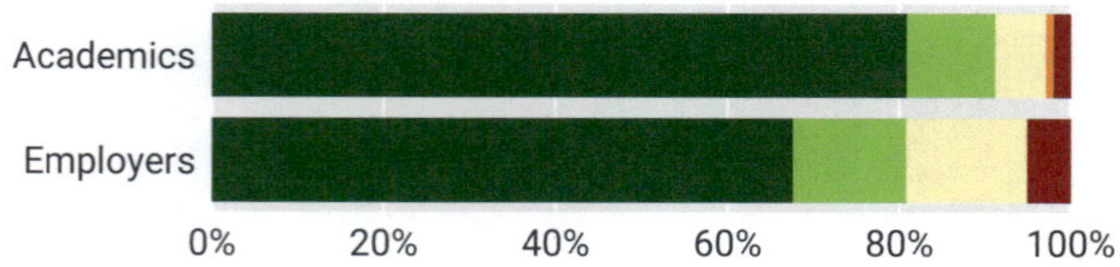

Figure A-5: Earth as a Complex System

Survey Category: Earth as complex and dynamic system with linkages between the different systems (e.g., lithosphere, atmosphere, biosphere, etc.)

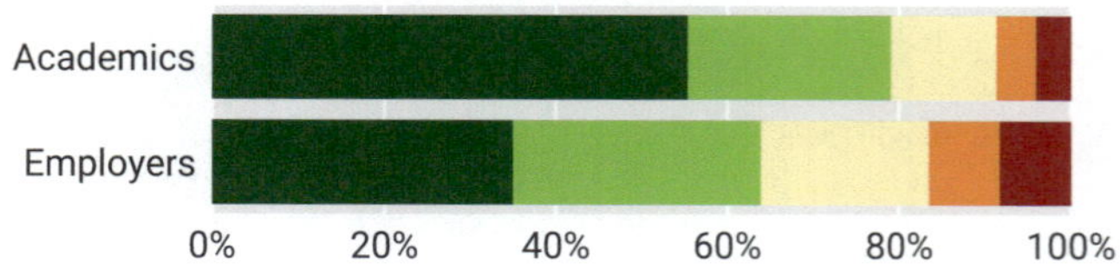

Figure A-6: Deep Time

Survey Category: Deep time (including the origin and evolution of life)

Continued on page 123

RESPONDENTS' RANKING OF IMPORTANCE OF TEACHING EACH SPECIFIC KEY GEOSCIENCE CONCEPT

1 Very important 2 3 4 5 Not important

Importance of Teaching Key Geoscience Concepts, continued from page 122

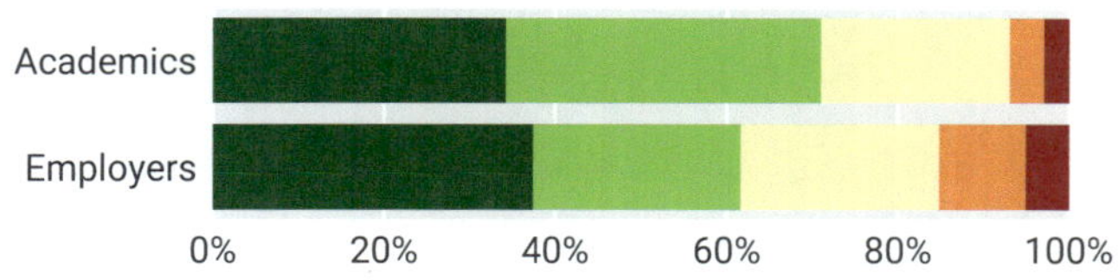

Figure A-7: Earth Materials

Survey Category: Earth materials

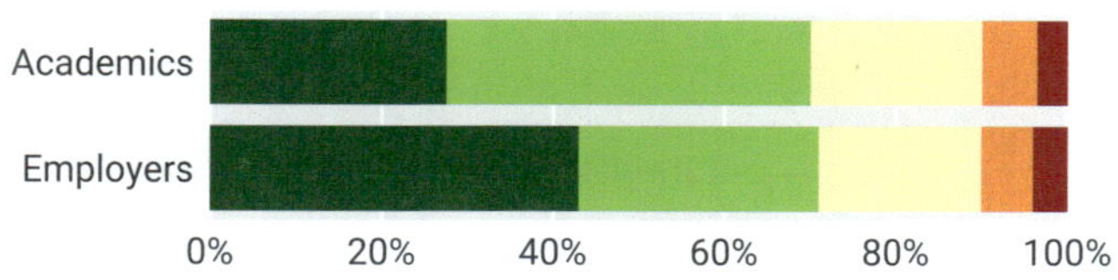

Figure A-8: Earth Structure

Survey Category: Earth structure

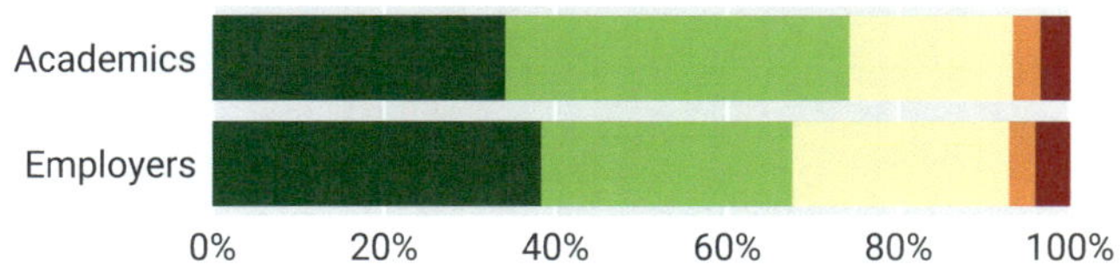

Figure A-9: Surface Processes

Survey Category: Surface processes (including relationship between landscape and process)

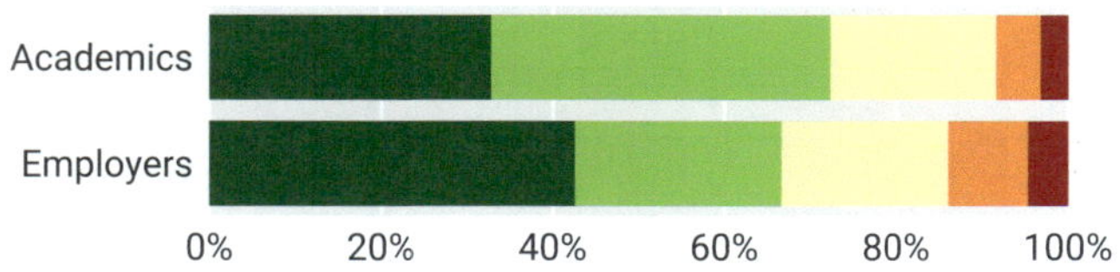

Figure A-10: Natural Resources

Survey Category: Natural resources (including energy)

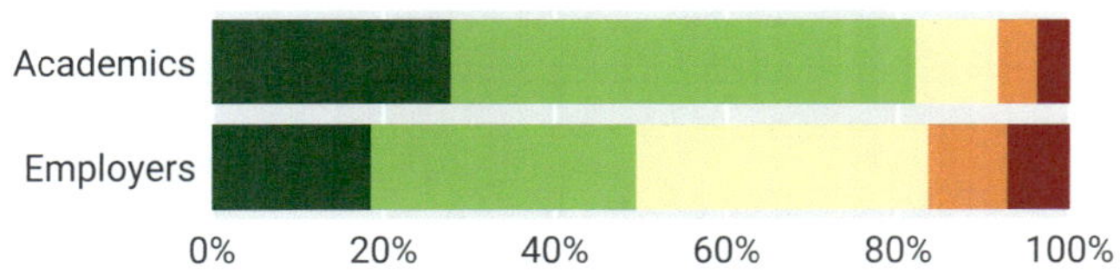

Figure A-11: Natural Hazards

Survey Category: Natural hazards

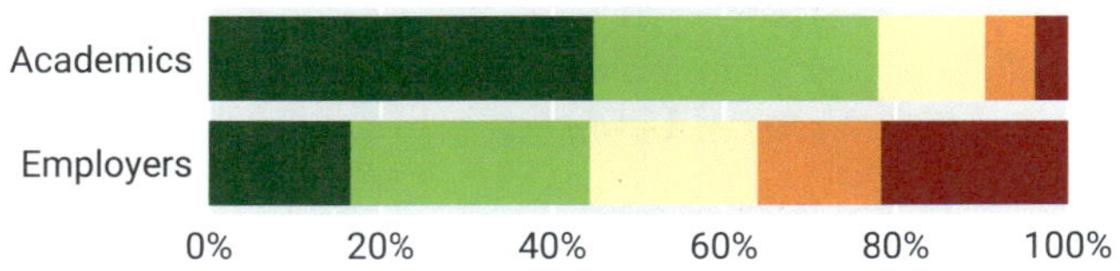

Figure A-12: Climate Change

Survey Category: Climate change

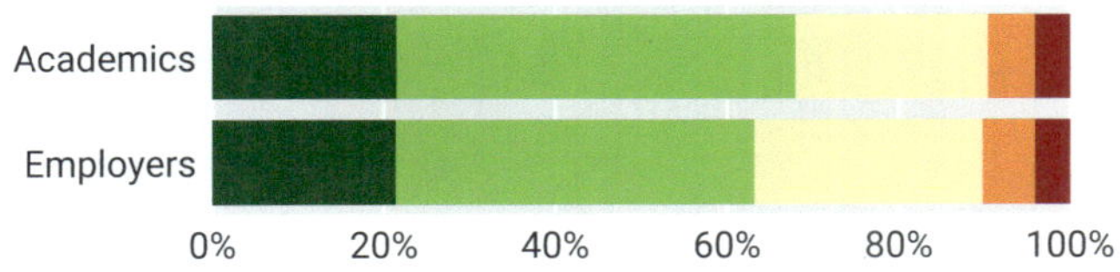

Figure A-13: Hydrogeology

Survey Category: Hydrogeology (including water, rock, and microbe interactions)

RESPONDENTS' RANKING OF IMPORTANCE OF TEACHING EACH SPECIFIC KEY GEOSCIENCE CONCEPT

1 Very important | 2 | 3 | 4 | 5 Not important

CRITICAL UNDERGRADUATE SKILLS AND COMPETENCIES

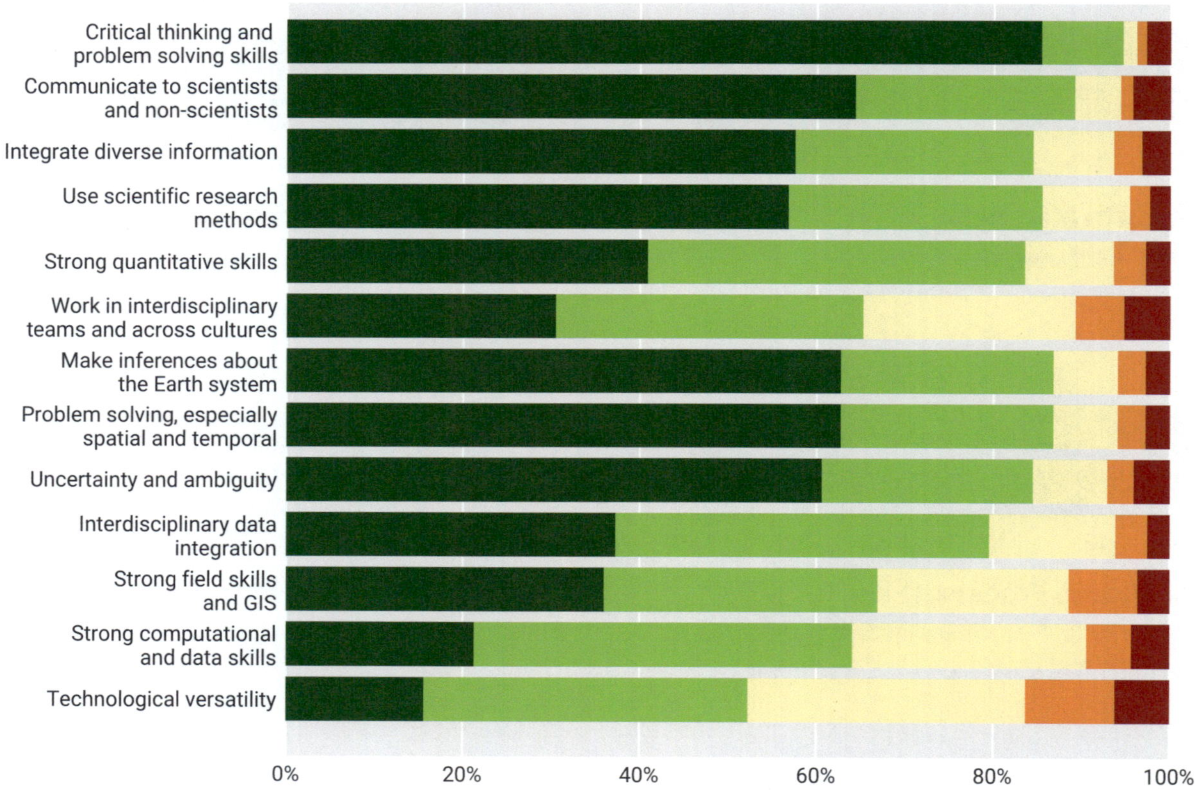

Figure A-14: Critical Undergraduate Skills and Competencies

Survey Question: What are the skills and competencies critical in undergraduate geoscience education?

Skills Importance Breakdowns by Respondent Employment Category

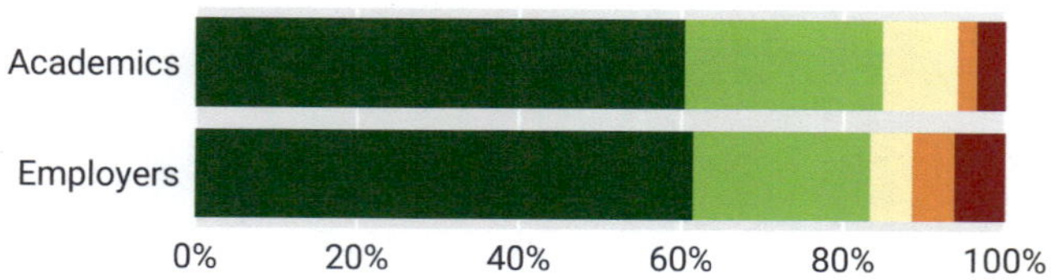

Figure A-15: Uncertainty and Ambiguity

Importance of skill

Survey Category: Work with uncertainty, non-uniqueness, incompleteness, ambiguity and indirect observations

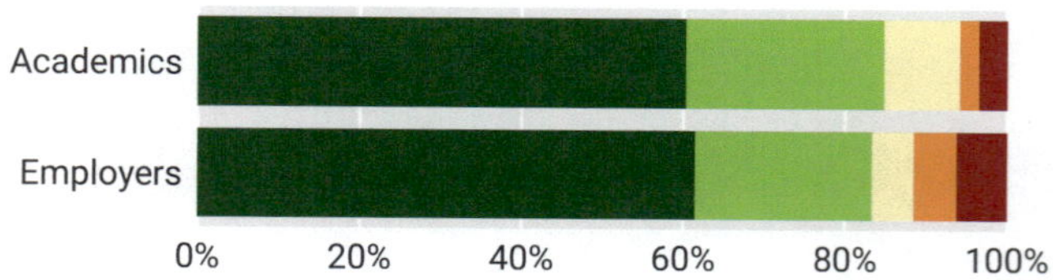

Figure A-16: Problem Solving

Importance of skill

Survey Category: Readily solve problems, especially those requiring spatial and temporal (i.e., 3D and 4D) interpretations

Continued on page 125

RESPONDENTS' RANKING OF SKILLS AND COMPETENCIES CRITICAL FOR UNDERGRADUATE STUDENTS TO LEARN

1 Very important — 2 — 3 — 4 — 5 Not important

Critical Undergraduate Skills and Competencies, continued from page 124

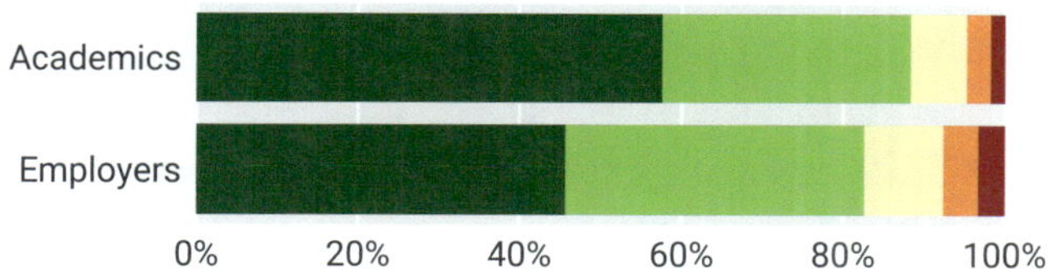

Figure A-17: Inferences about Earth System

Importance of skill

Survey Category: Make inferences about Earth system from observations of natural world combined with experimentation and modeling

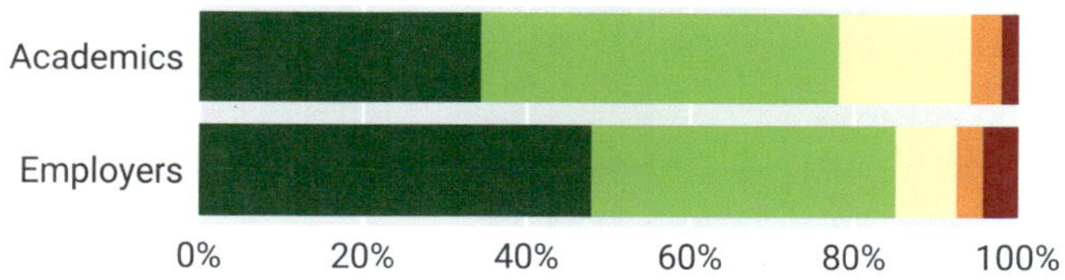

Figure A-18: Integrate Multidisciplinary Data

Importance of skill

Survey Category: Integrate data from different disciplines and apply systems thinking

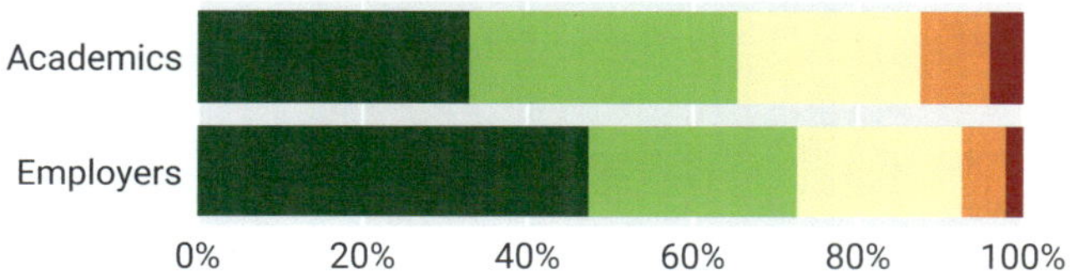

Figure A-19: Strong Field and GIS skills

Importance of skill

Survey Category: Have strong field skills and a working knowledge of GIS

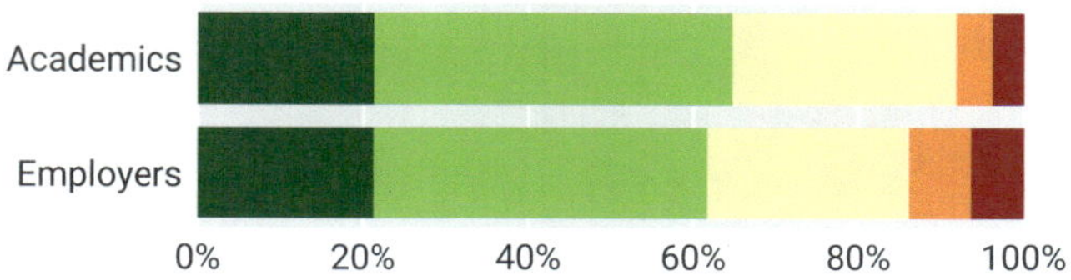

Figure A-20: Computational and Data Skills

Importance of skill

Survey Category: Have strong computational skills and the ability to manage and analyze large datasets

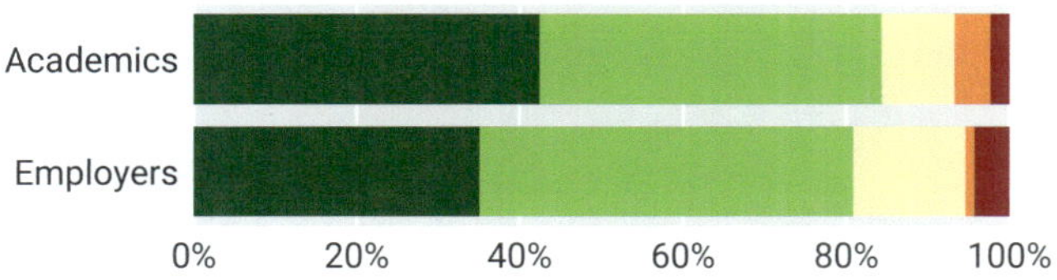

Figure A-21: Strong Quantitative Skills

Importance of skill

Survey Category: Have strong quantitative skills and ability to apply

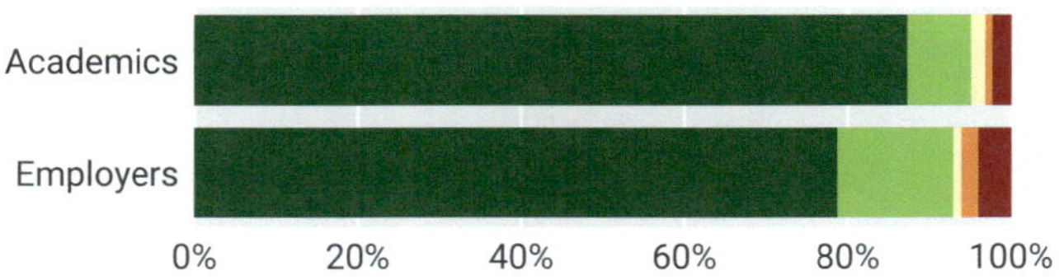

Figure A-22: Critical Thinking

Importance of skill

Survey Category: Critical thinking/problem solving skills

Continued on page 126

RESPONDENTS' RANKING OF SKILLS AND COMPETENCIES CRITICAL FOR UNDERGRADUATE STUDENTS TO LEARN

1 Very important — 2 — 3 — 4 — 5 Not important

Critical Undergraduate Skills and Competencies, continued from page 125

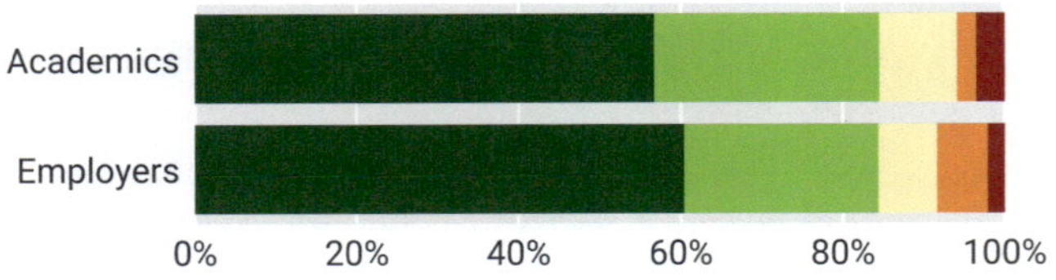

Figure A-23: Integrate Disparate Information

Importance of skill

Survey Category: Ability to access and integrate information from different sources and to continue to learn

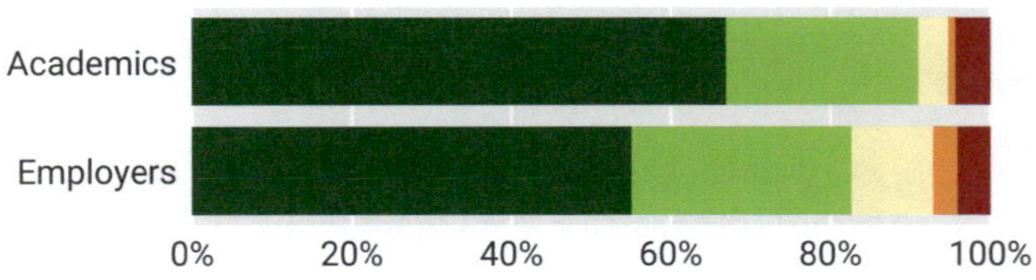

Figure A-24: Communicate Effectively

Importance of skill

Survey Category: Communicate effectively to scientists & non-scientists

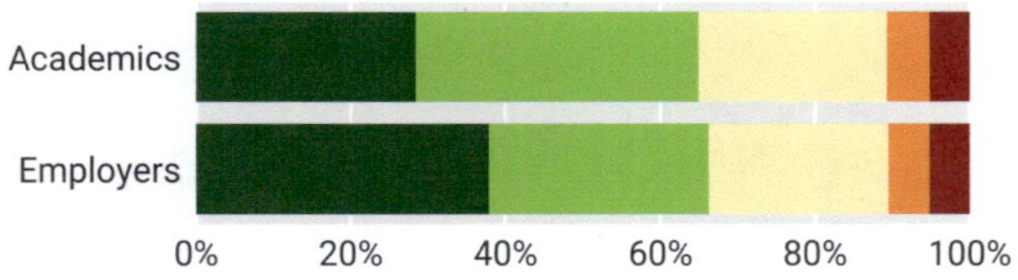

Figure A-25: Work in Teams and Across Cultures

Importance of skill

Survey Category: Work in interdisciplinary teams and across cultures

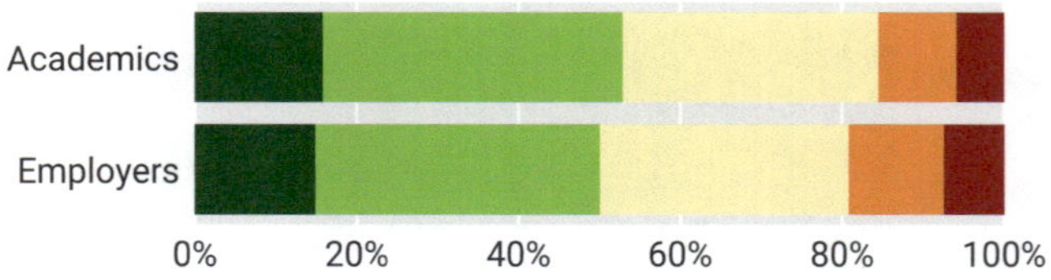

Figure A-26: Technological Flexibility

Importance of skill

Survey Category: Be technologically versatile (i.e., Google Earth, tablets, smartphones, apps)

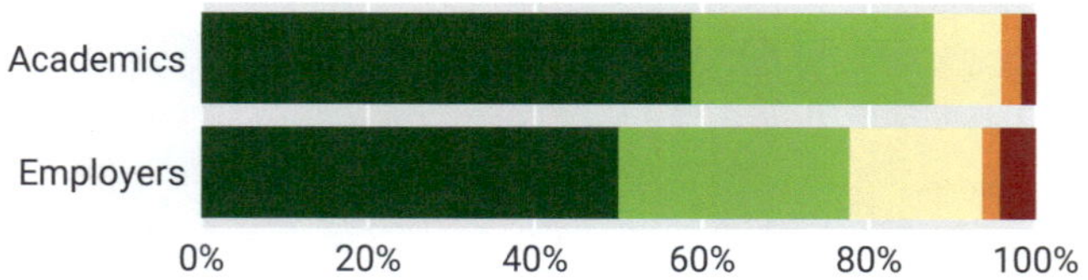

Figure A-27: Scientific Research Methods

Importance of skill

Survey Category: Understand and use scientific research methods

RESPONDENTS' RANKING OF SKILLS AND COMPETENCIES CRITICAL FOR UNDERGRADUATE STUDENTS TO LEARN

1 Very important | 2 | 3 | 4 | 5 Not important

ASSESSMENT OF DEPARTMENTAL VIEWS

Is your department interested in making changes to your undergraduate curriculum to focus on competencies, skills, and conceptual understanding?

Is the development of your institution's undergraduate curriculum, in terms of competencies, skills, or conceptual understanding, regulated at any institutional, local, or state levels?

Does your institution track Student Learning Outcomes or other metrics of student success with in your department?

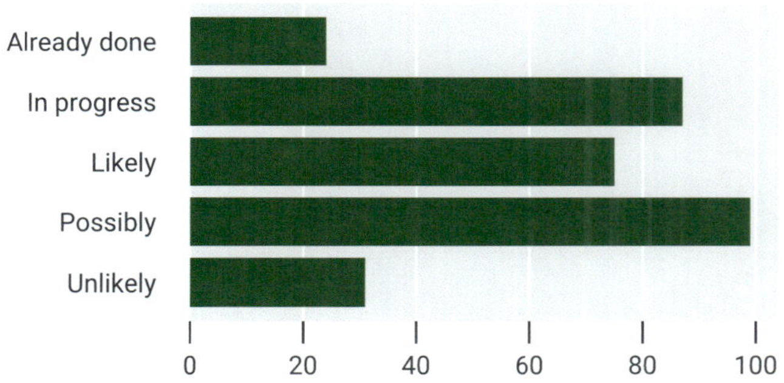

Figure A-28: Focus on Competencies

Departments reporting

Survey Question: How likely is it that your department will make a systematic effort to improve competencies, skills, and conceptual understanding in developing your undergraduate curriculum in the next few years?

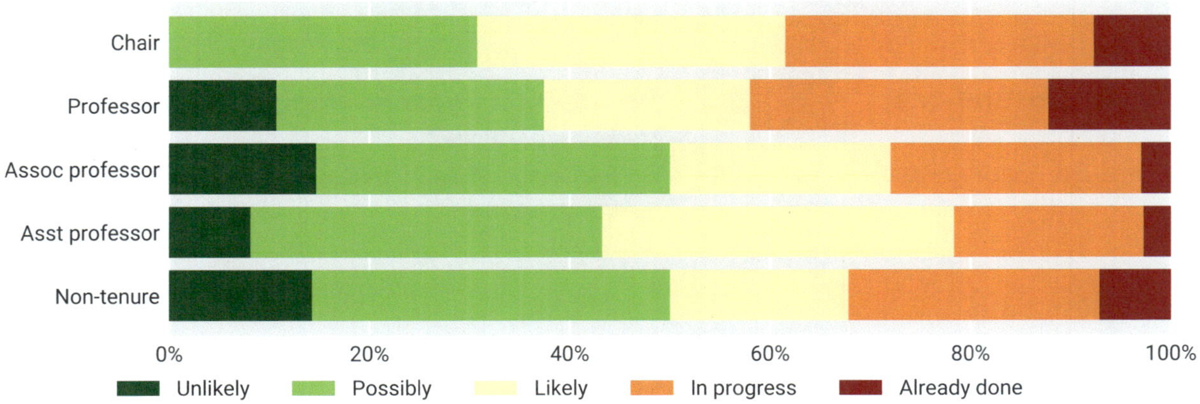

Figure A-29: Likelihood of Systemic Reform of Undergraduate Curriculum

Respondents by academic rank

Survey Question: How likely is it that your department will make systematic efforts to encourage faculty to incorporate research-validated teaching strategies?

OPPORTUNITIES AND ACTIVITIES AVAILABLE TO UNDERGRADUATES

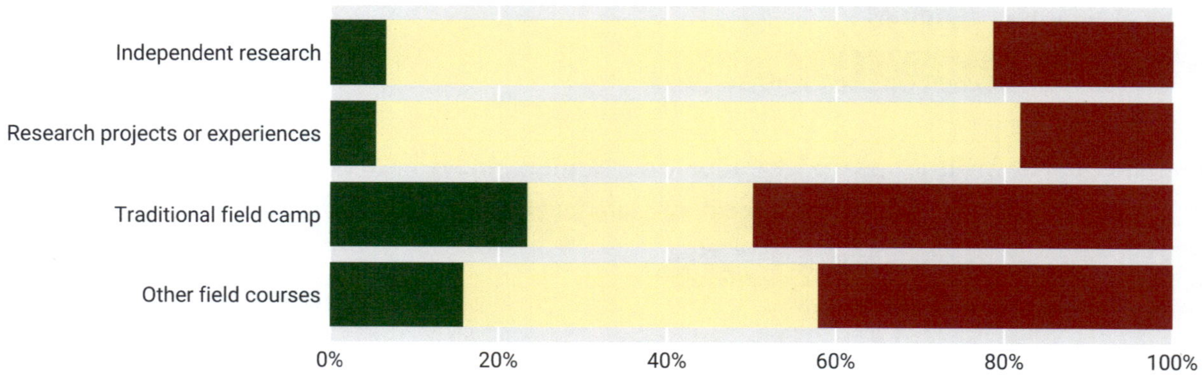

Figure A-30: Undergraduate Opportunities

Departments reporting

Survey Questions: Do your undergraduates have the opportunity for the following activities?

Breakdowns of Activity Availability by Institution Type

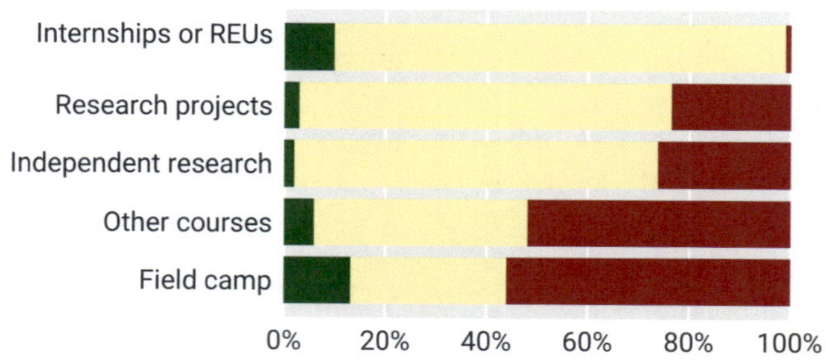

Figure A-31: 4-year Public Institutions

Student opportunities

Survey Category: Undergraduate opportunities at 4-year public institutions

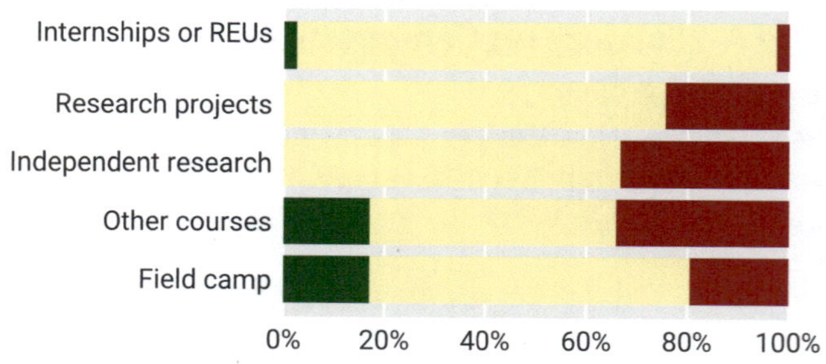

Figure A-32: 4-year Private Institutions

Student opportunities

Survey Category: Undergraduate opportunities at 4-year private institutions

Continued on page 129

RESPONDENTS' REPORTING OF OPPORTUNITIES AND ACTIVITIES AVAILABLE TO UNDERGRADUATES AT THEIR INSTITUTIONS

Opportunities and Activities Available to Undergraduates, continued from page 128

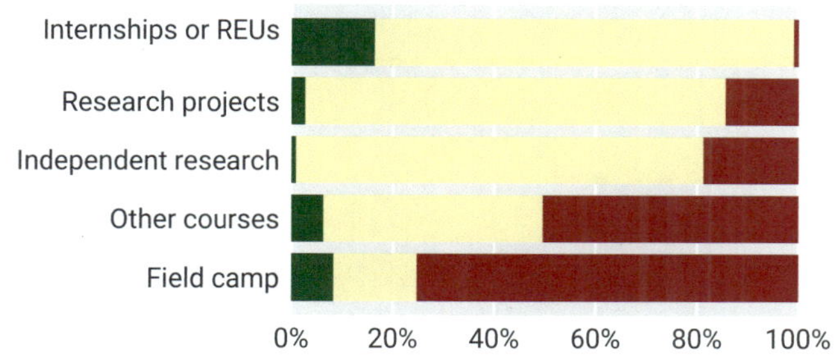

Figure A-33: Public Research Institutions

Student opportunities

Survey Category: Undergraduate opportuntities at public research institutions

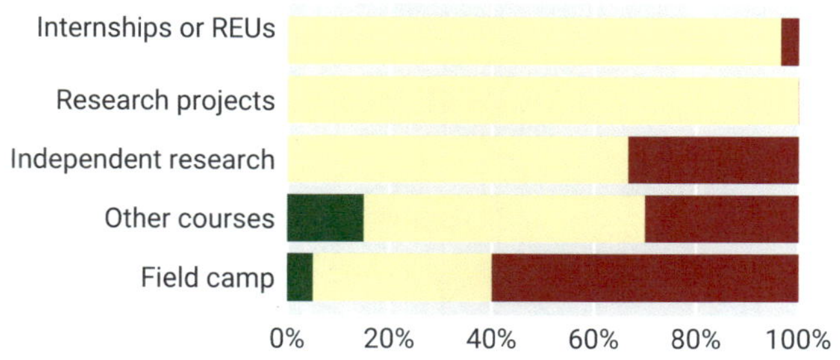

Figure A-34: Private Research Institutions

Student opportunities

Survey Category: Undergraduate opportunities at private research institutions

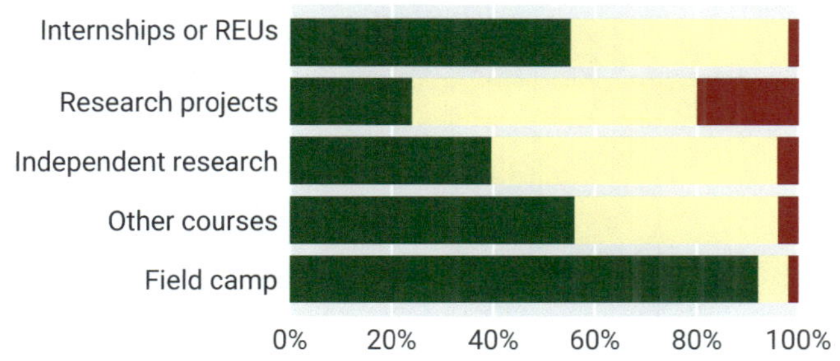

Figure A-35: 2-year Colleges

Student opportunities

Survey Category: Undergraduate opportunities at 2-year colleges

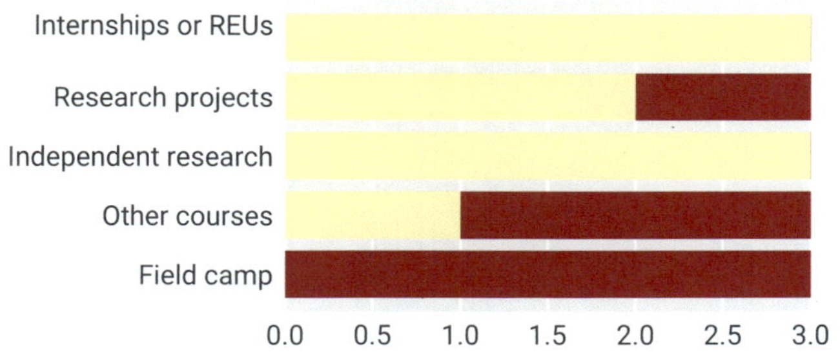

Figure A-36: Hispanic-Serving Institutions

Number of institutions

Survey Category: Undergraduate opportunities at Hispanic-serving institutions

RESPONDENTS' REPORTING OF OPPORTUNITIES AND ACTIVITIES AVAILABLE TO UNDERGRADUATES AT THEIR INSTITUTIONS

TEACHING METHODS USED BY FACULTY

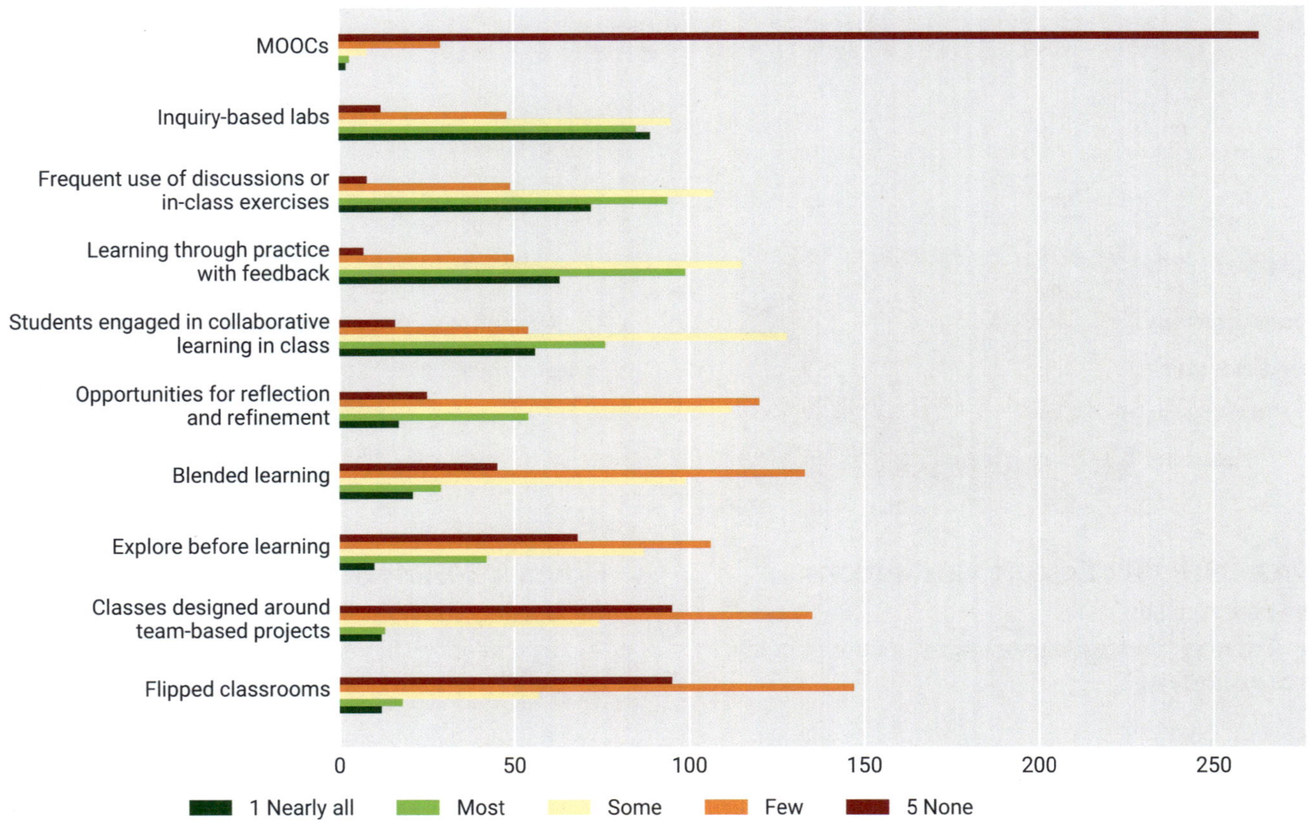

Figure A-37: Teaching Methods Used by Faculty

Departments reporting

Survey Question: From what you know about your department, which of the below teaching methods are used by faculty in your department?

Continued on page 131

Teaching Methods Used by Faculty, continued from page 130

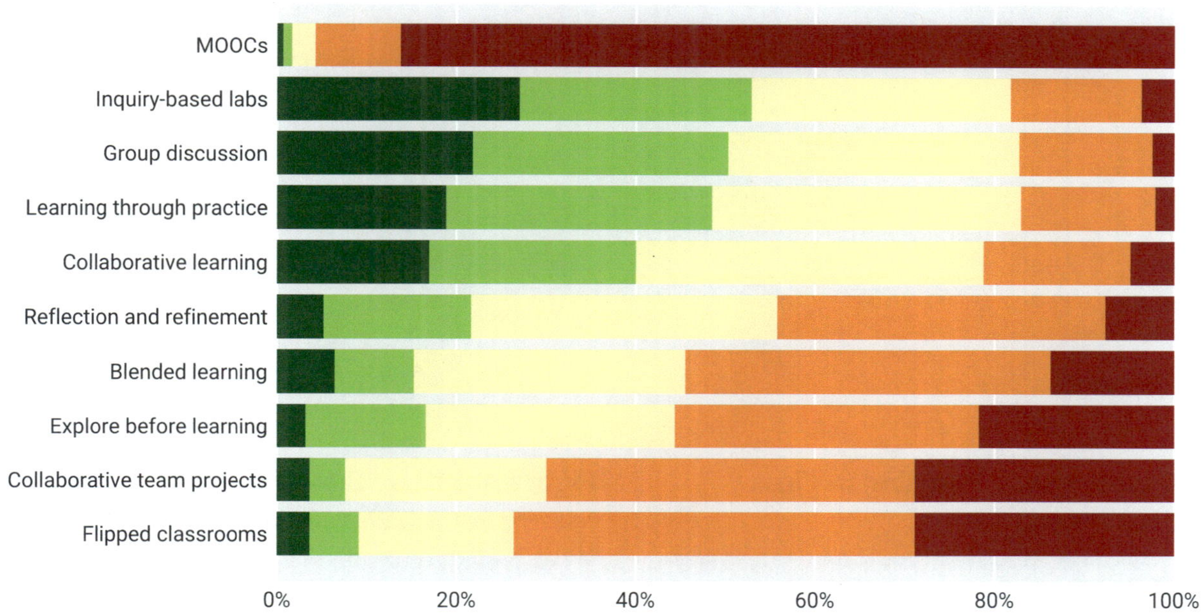

Figure A-38: Teaching Methods Used

Percentage of departments

Survey Question: From what you know about your department, which of the below teaching methods are used by faculty in your department?

Breakdowns of Teaching Methods by Institution Type

Notation: 2YC is 2-Year College; 4YC is 4-Year College; RU is Research University

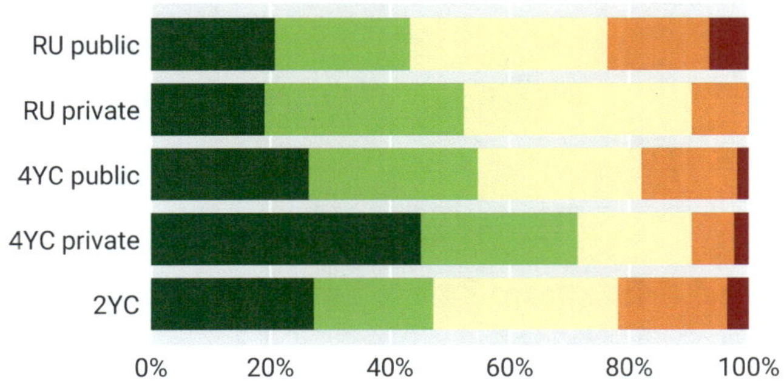

Figure A-39: Inquiry-based Labs

Teaching method by institution type

Survey Category: Inquiry-based labs

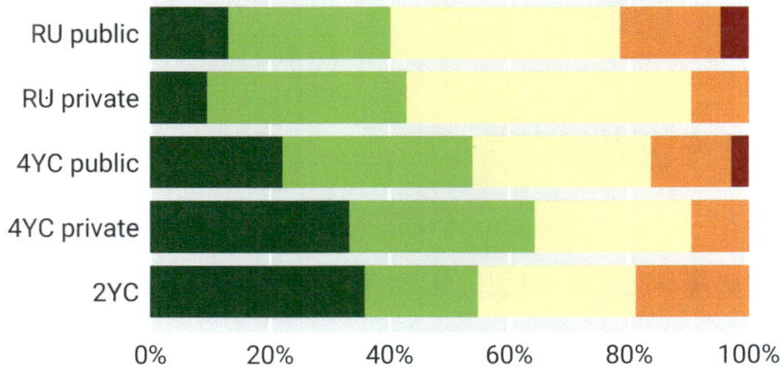

Figure A-40: Frequent Use of Discussions

Teaching method by institution type

Survey Category: Frequent use of small group discussion, whole class discussion or in-class exercises

Continued on page 132

RESPONDENTS' REPORTING OF TEACHING METHODS USED IN THEIR DEPARTMENTS

Teaching Methods Used by Faculty, continued from page 131

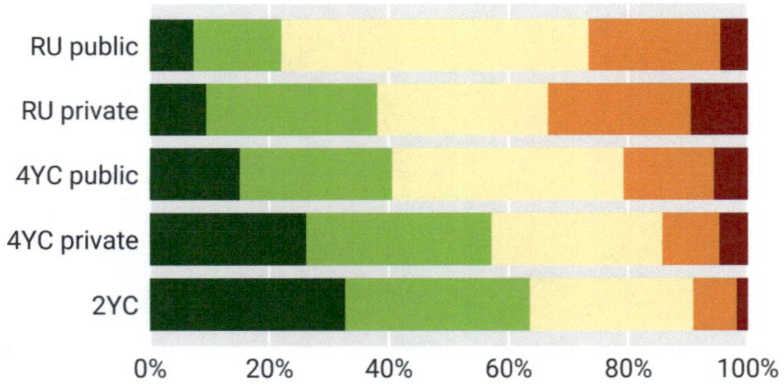

Figure A-41: Collaborative Learning in Class

Teaching method by institution type

Survey Category: Students engaged in collaborative learning in class (e.g., Think-Pair-Share; team exercises/discussions)

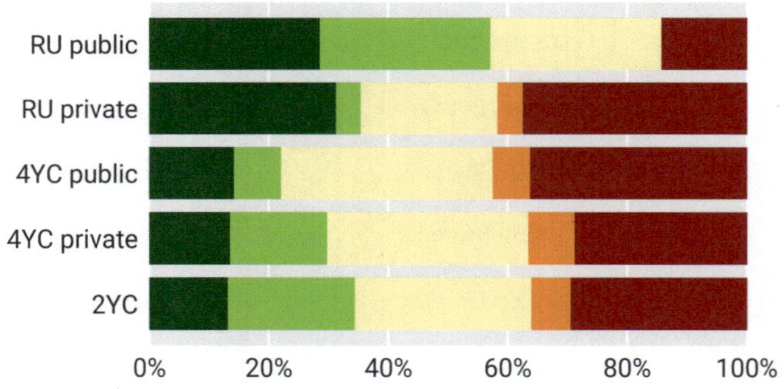

Figure A-42: Use Real Data & Research

Teaching method by institution type

Survey Category: Learning through practice with feedback — teaching with using real data & research

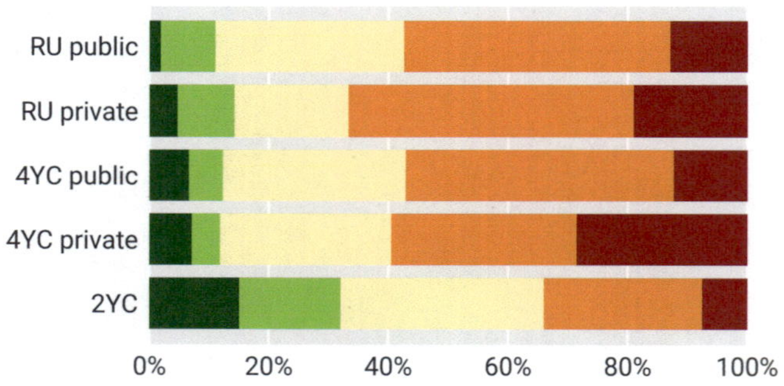

Figure A-43: Blended Learning

Teaching method by institution type

Survey Catgory: Blended learning (classroom lectures/ activities combined with partial online delivery of content and instruction)

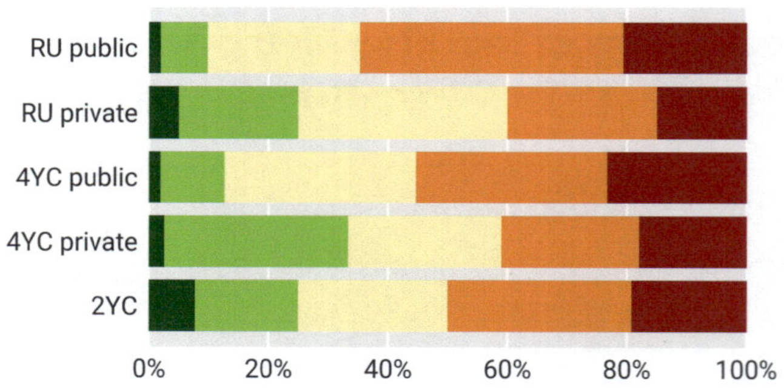

Figure A-44: Explore Before Learning

Teaching method by institution type

Survey Category: Explore before learning

Continued on page 133

RESPONDENTS' REPORTING OF TEACHING METHODS USED IN THEIR DEPARTMENTS

5 All | 4 | 3 | 2 | 1 None

Teaching Methods Used by Faculty, continued from page 132

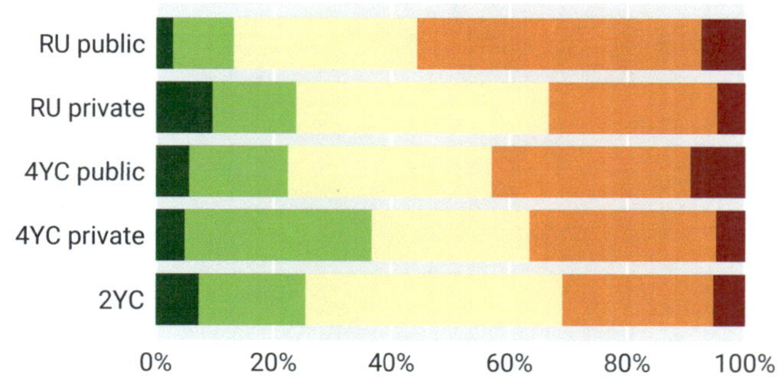

Figure A-45: Reflection and Refinement

Teaching method by institution type

Survey Category: Opportunities for reflection and refinement (e.g., Retrieval practice, Minute papers, Concept tests, etc.)

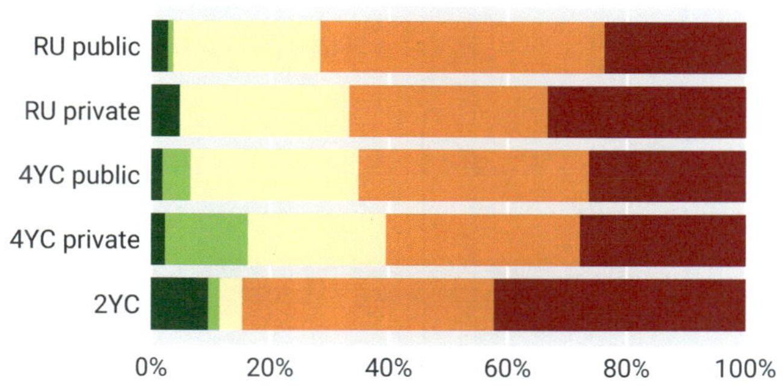

Figure A-46: Collaborative Team Projects

Teaching method by institution type

Survey Category: Entire classes designed around collaborative team based projects

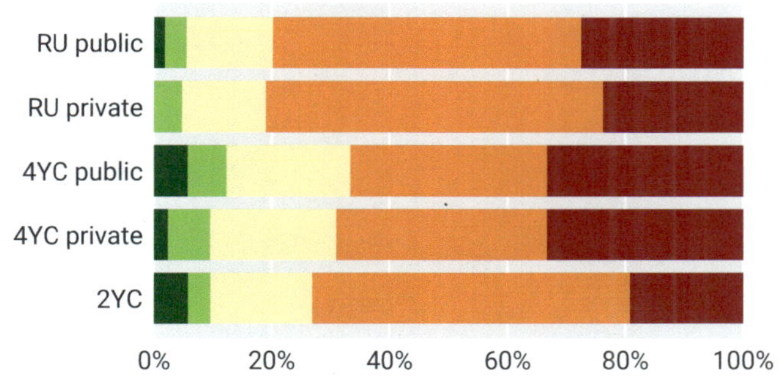

Figure A-47: Flipped Classroom

Teaching method by institution type

Survey Category: Flipped classroom

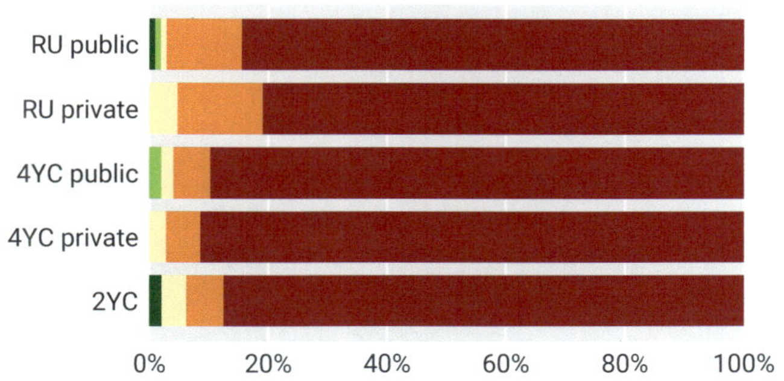

Figure A-48: MOOCs

Teaching method by institution type

Survey Category: MOOCs

RESPONDENTS' REPORTING OF TEACHING METHODS USED IN THEIR DEPARTMENTS

5 All | 4 | 3 | 2 | 1 None

TECHNOLOGY USED IN TEACHING

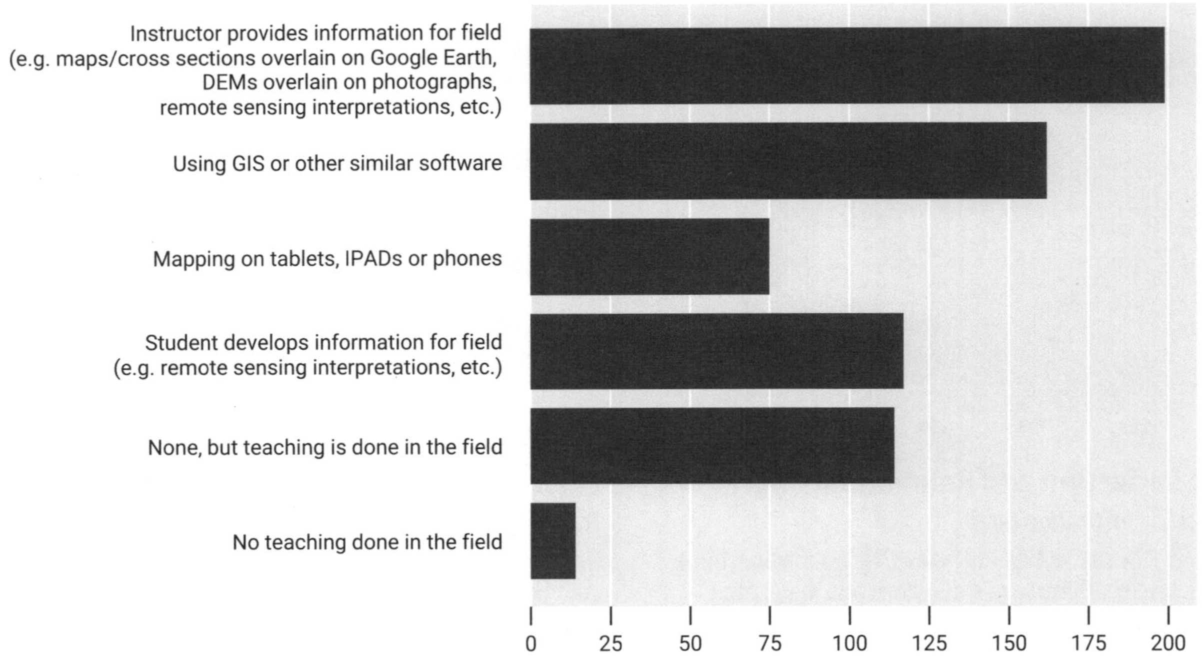

Figure A-49: Technology Used in Field Teaching

Departments reporting

Survey Question: From what you know about your department, how do your colleagues (or you) use technology in teaching in the field?

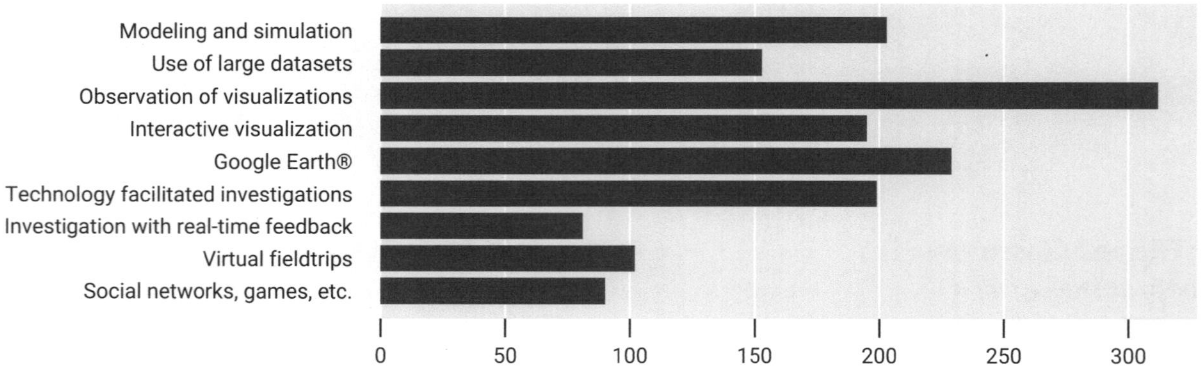

Figure A-50: Technologies Used in Teaching

Departments reporting

Survey Question: From what you know about your department, how do your colleagues (or you) use technology in teaching?

USE OF VALIDATED TEACHING STRATEGIES

Is your department interested in making changes to how teaching is done at the undergraduate level?

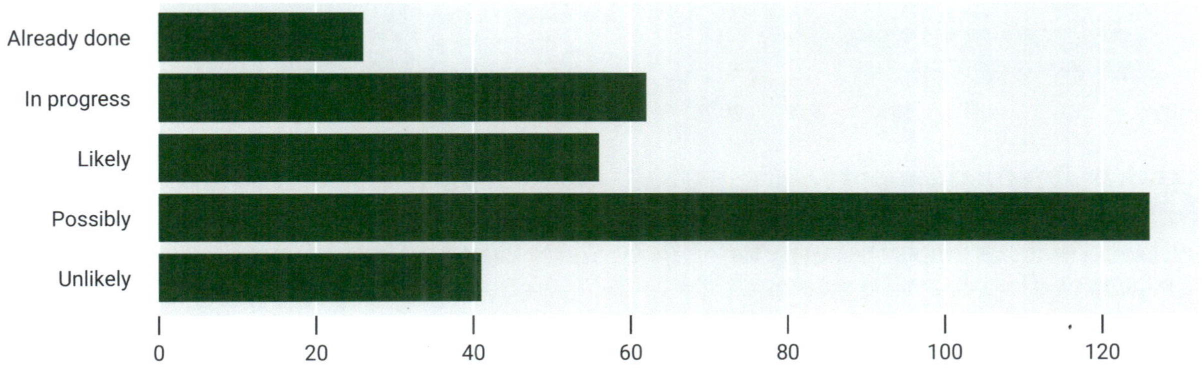

Figure A-51: Likelihood of Using Research-Validated Teaching Strategies

Departments reporting

Survey Question: How likely is it that your department will make systematic efforts to encourage faculty to incorporate research-validated teaching strategies?

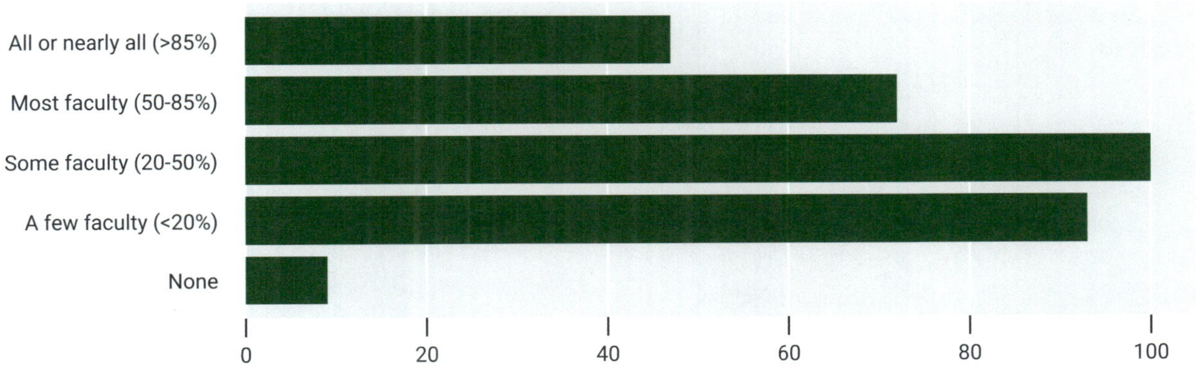

Figure A-52: Faculty Using Reformed Pedagogy

Departments reporting

Survey Question: How extensive are efforts to introduce pedagogical reforms in your department?

OBSTACLES/BARRIERS IDENTIFIED BY SUMMIT TO IMPLEMENTING RESEARCH-VALIDATED PEDAGOGIES AND USES OF TECHNOLOGY

Breakdowns of Impacts by Academic Rank

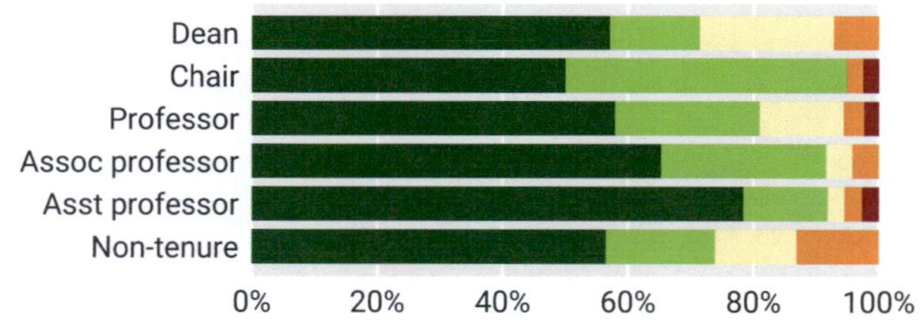

Figure A-53: Lack of Time and Support

Impact by academic rank

Survey Category: Lack of time and support necessary for developing and piloting new instructional approaches

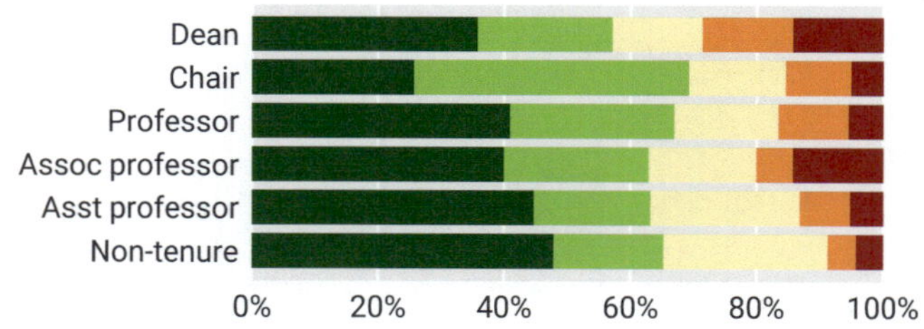

Figure A-54: Financial Resources

Impact by academic rank

Survey Category: Financial resources

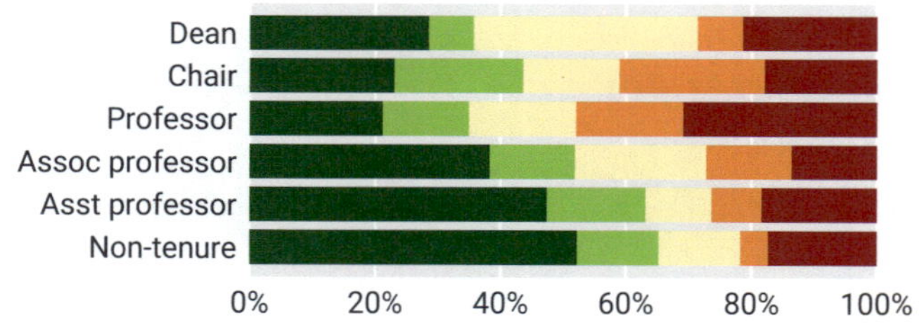

Figure A-55: Performance and T&P Reviews

Impact by academic rank

Survey Category: Annual performance and tenure and promotion evaluations

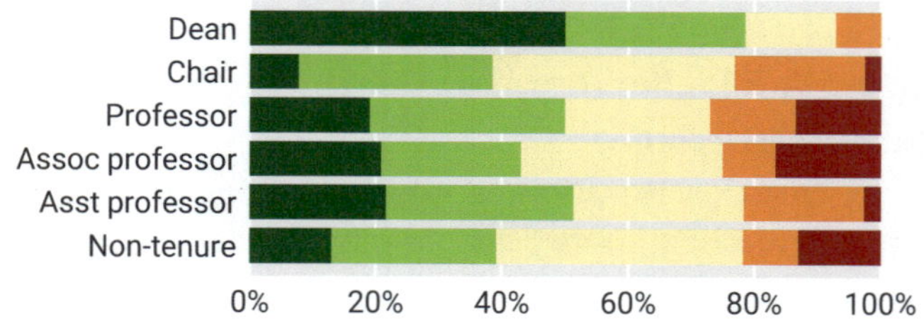

Figure A-56: Need Validated Techniques

Impact by academic rank

Survey Category: Lack of information on what techniques are research-validated

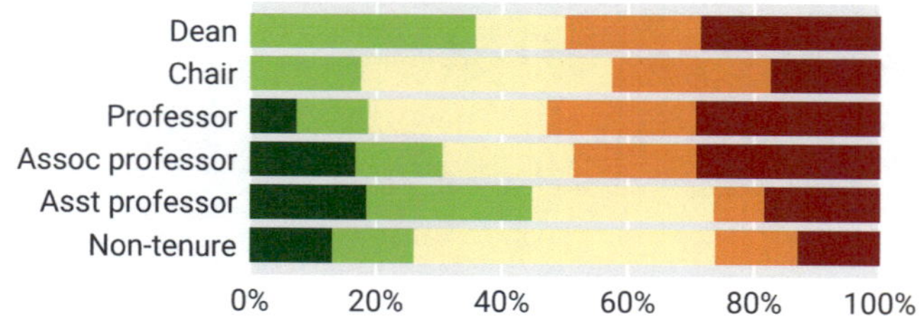

Figure A-57: Student Evaluations

Impact by academic rank

Survey Category: Concern about student evaluations

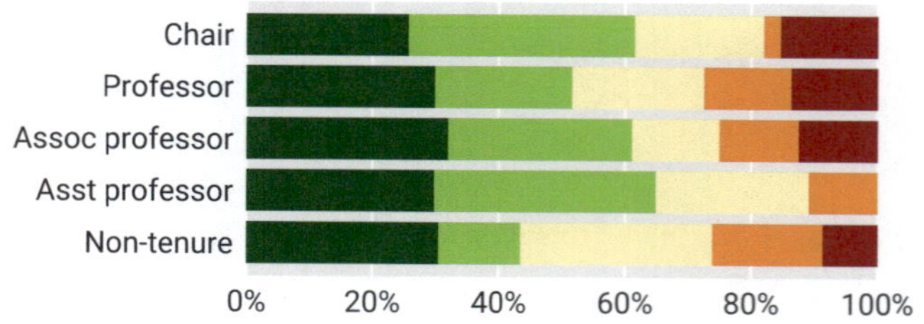

Figure A-58: Instructional Infrastructure

Impact by academic rank

Survey Category: Instructional space design and teaching infrastructure

SUMMIT PARTICIPANTS' ASSESSMENTS OF BARRIERS TO IMPLEMENTING RESEARCH-VALIDATED PEDAGOGIES AND TECHNOLOGY USE

PROFESSIONAL DEVELOPMENT AND K–12 TEACHER TRAINING

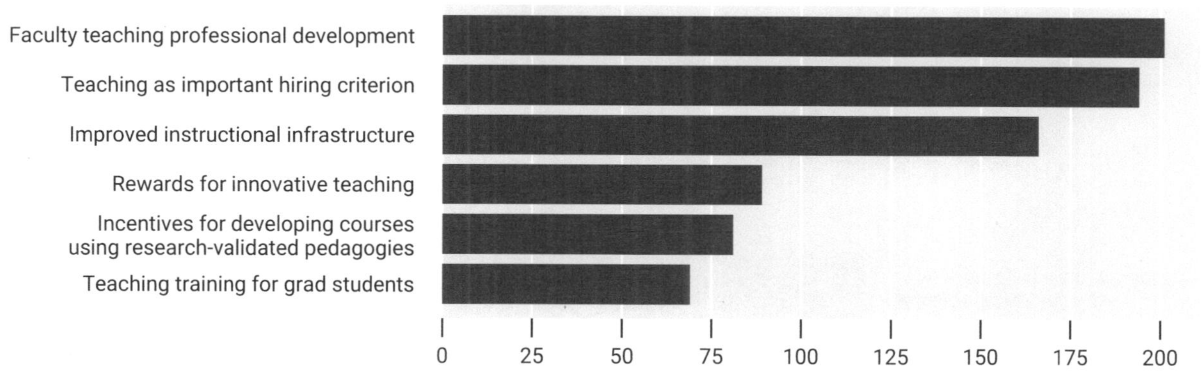

Figure A-59: Teaching Development Opportunities

Departments reporting

Survey Question: Does your department use or offer any of the following?

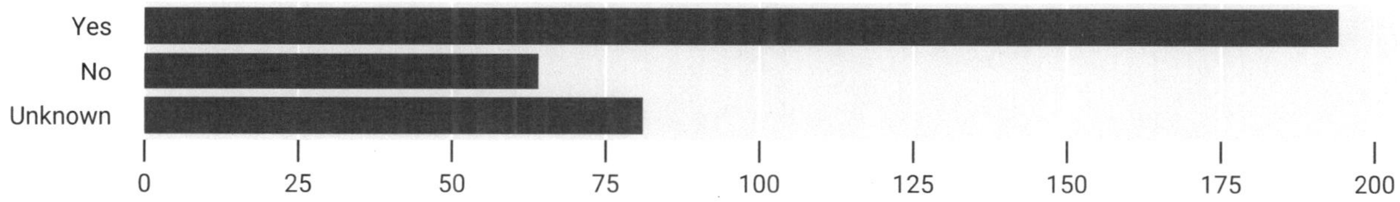

Figure A-60: Is Earth Science Taught in Local K-12 schools?

Departments reporting

Survey Question: Is Earth Science (geoscience) taught as a course or in the curriculum in K-12 in your local school districts?

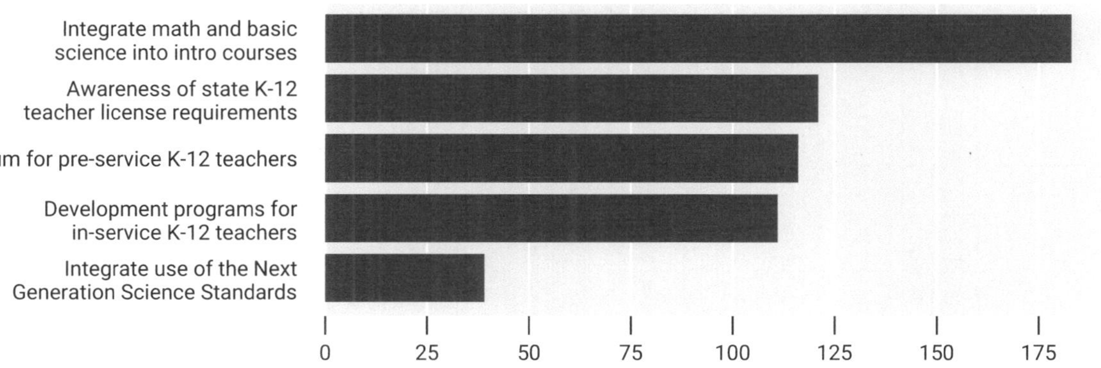

Figure A-61: Ways Department Helps K-12 Teacher Training

Departments reporting

Survey Question: Which of the following does your department do to help with preparation of K-12 teachers?

Continued on page 138

Professional Development and K–12 Teacher Training, continued from page 137

Does your department/company/organization sponsor, partner with or have any professional development programs for in-service K–12 teachers?

Yes: 156 **No: 272**

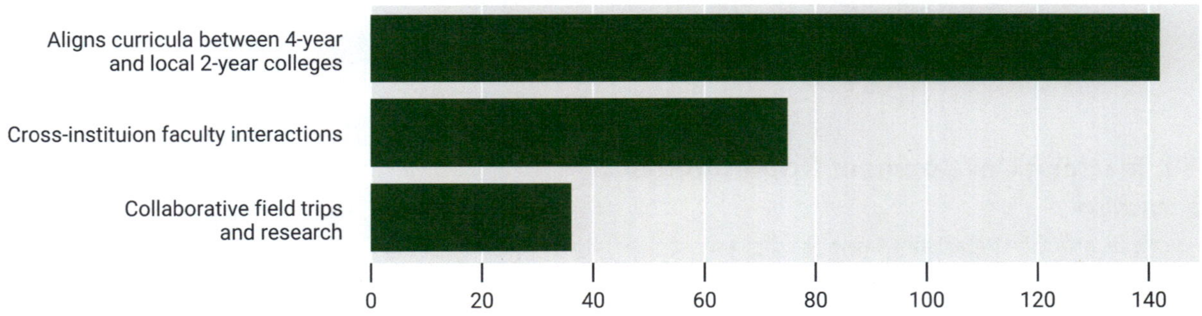

Figure A-62: Programs for 2-year to 4-year College Transition

Departments reporting

Survey Question: Which of the following does your department do to ease the transition between 2-year and 4-year colleges?

EFFORTS TOWARD STUDENT DIVERSITY AND BROAD PARTICIPATION

Does your department/company/organization have or plan on any systematic efforts to encourage broadening participation and retention of a more diverse student population?

Yes: 181 **No: 239**

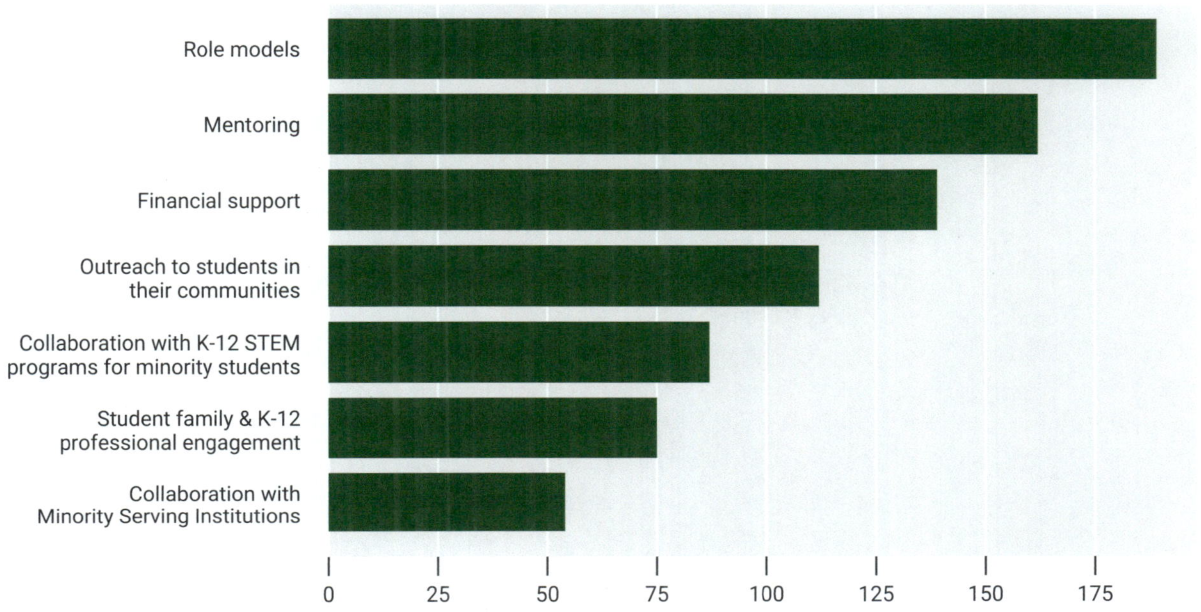

Figure A-63: Systematic Efforts to Broadening Participation

Organizations reporting

Survey Question: Does your department/company/organization have or plan on any systematic efforts to encourage broadening participation and retention of a more diverse student population?

Does your department/company/organization track the participation and retention of minorities in your population?

Yes: 237 **No: 182**

© Michael Collier

Appendix B: Geoscience Employers Workshop Outcomes

A **Geoscience Employers Workshop** was held May 27–28, 2016 in Washington D.C. to get employers of geoscientists' input on the developing community vision for the geosciences. The 46 participants were evenly distributed between the petroleum industries; hydrology, engineering and environmental consulting companies; and federal agencies that employ geoscientists, along with representatives from some of the geoscience professional societies. One participant represented the mining.

The first breakout session asked participants to identify the skills, competencies, and conceptual understandings needed by geoscience bachelor's recipients for future employment. Subsequently the results from the Summit and survey were presented. A series of breakout sessions compared the employers' initial views and that of the results from the Summit and survey. Overall, agreement among the employers and between their views and the results of the developing vision from the Summit and survey was remarkably strong. In addition to their own views on general skills and competencies, they also provided detailed granularity to the concepts and skills identified by the Summit. Lastly, they discussed ways to implement or develop these skills and competencies in students, the role industry and other employers should take to help the academic community, and how to develop better academic-industry partnerships.

The specific results are summarized below.

IMPORTANT CONCEPTS IDENTIFIED BY THE EMPLOYERS

Systems Thinking: How Systems Work and Interact

- Atmosphere — climate, weather, ocean-atmospheric circulation
- Hydrosphere — ocean, ice, surface water, groundwater
- Lithosphere — rock cycle, deformation, structure, tectonics
- Pedosphere/surface — geomorphic, erosion, and surface processes, landscape evolution
- Biosphere — paleontology, ecosystems
- Solar/Earth interactions — tidal, climate, planetary geology
- Human/societal coupled to Earth — natural resources, energy, anthropomorphic climate change, natural hazards
    - Influence of geology on society
    - Influences of society on earth processes

Processes

- Geochemical cycles — C, H_2O, N, P
- Thermodynamics — energy, kinetics, diffusion, heat, mass transfer, fluid flow
- Geomechanics/stress state/rheology
- Geological time/Earth evolution
- Plate tectonics/geodynamics
- Tectonic processes
- Depositional processes
- Crystallization processes

Tools

- Statistics/uncertainty/probability
- Mathematics (differential equations, linear algebra)
- Cartography
- Geography and spatial thinking
- Field methods
- Potential fields
- Remote sensing

- Age dating
- Instrumentation
- Analytical/numerical modeling
- Seismology/geophysical sensing

The employers also provided more granularity for each of the general concepts identified by the Summit.

Earth as a Complex System

- Nonlinear complex systems
  - Size of systems — complexity of scale and interactions
  - Feedback loops, interactions, forcings
  - Implications and predictions
- Energy, mass, fluid transport (movement and flow), residency, and cycles
- Work/changes that affect the Earth's systems
  - Human drivers and impacts of change, Anthropocene
  - Environmental transitions
  - Scales of change
  - Using the present processes to infer past processes: Advantages/risks
- Solar system interaction

Deep Time

- Conventional concepts of geologic time
  - Paleontology, superposition
  - Relative vs. absolute age
  - Tools to determine absolute age (radioisotopes, stable isotopes, etc.), precision of data, limitations
  - Extrapolate from lab to field
- Impact on processes
  - Time scales over which processes are relevant
  - Specific periods in geologic time that are critical for different processes
  - Impact of time on "Earth" events (i.e., weathering, geodynamics, resources, etc.)
- Events and rates
  - Duration, frequency, magnitude and residence time
  - Timing, scale, sequencing, and rates of change
- Temporal reasoning

Climate Change

- What is climate change? — geologic scale vs. present change
  - Significant climate change in geologic past
  - Relevant space and time scales
  - Continental vs. local scale change
  - Proxy records
  - Rate of climate change; rapid change
- Driving forces and causal mechanisms
  - External forcing vs. internal forcing
  - Dependence upon spatial and temporal scale and feedbacks
  - Impact of plate tectonics, atmosphere-earth interactions, etc.
  - Human-induced climate change
- Carbon cycle
- Difference between weather and climate
- Impacts of climate change
  - Water resources, hydrologic cycle, other climate-change effects
  - Biosphere implications, ocean acidification, sea-level rise
  - Implications on soil, agriculture
  - Economics and social aspects of climate change
  - Climate element to environmental consulting and hydrogeology as well as petroleum exploration

Natural Resources

- Understanding of what is included in "natural resources"
  - Economic geology (commodities and finite resources)
  - Energy, water, minerals, geologic materials
- Solid vs. liquid resources, geographic distribution, uses
- Ecosystem services, analysis of renewable and non-renewable (finite) resources
- Resource dependency and limits
  - Finite resource or commodity
  - Understanding your environment (where do our materials, energy, and medicines come from)
  - Ore and fossil fuel supply and demand and getting it to market
  - Time and space scale of formation and depletion, sustainability
  - Economics and viability of resources
  - How things are made

- Process from ore to refined product
- Process from fossil fuel to energy or material objects

Surface Processes

- Sediment deposition & erosion
  - Stream/River flow, morphology, deposition, erosion, effect of floods
  - Transport relationships (all surface processes)
  - Magnitude and frequency relationships of surficial deposits
  - Subsurface analogs
- Terrestrial and marine surface interactions
  - Biological, chemical, and physical interactions
  - Rates of chemical and physical changes
- Landscape alteration (geomorphology)
  - Surface mechanical processes
  - Karst formation
  - Glacial till and overburden thickness
- Habitability, sustaining life
  - Ties to natural hazards

Earth Materials

- What is a rock, mineral? Rock cycle
- Rocks: physical and chemical properties
  - How measure, scale of measurement
  - Mechanical characteristics
  - Scales of heterogeneity
  - How change over time
- Processes that form rocks and minerals
  - Processes and conditions of formation
  - Localizing mechanisms for deposits
  - Fluid dynamics, flow and fluid chemistry
  - Role of microbiology and organisms
- Resource applications, organic-inorganic materials

Earth Structure

- Structure of Earth
  - Mechanical and compositional layers
  - Tools for defining earth structure (seismic waves, analysis of earthquakes, etc.)
- Deformation
  - Stress and strain
  - Rock mechanics and deformation processes
  - Fractures, faults, folds, other structural features, etc.
- Plate Tectonics, including
  - Basin formation
  - Episodic nature, planning perspectives, uncertainty
- Structural controls on resource accumulations

Hydrogeology

- Water cycle
- Groundwater/aquifers, confined vs. unconfined aquifers
  - Phase behaviors
  - Saturated vs. unsaturated conditions
  - Scales of heterogeneity in space and time
  - Contaminant transfer
- Biogeochemistry and aqueous geochemistry
  - Microbe interactions
  - Nutrient cycling
- Subsurface-surface interactions
- Economics and public policy
  - Groundwater quality
  - Regulatory standards

Courtesy of the Jackson School of Geosciences, University of Texas at Austin

IMPORTANT SKILLS AND COMPETENCIES IDENTIFIED BY THE EMPLOYERS

Geoscience Thinking

- Earth science habits of mind/geoscientific thinking
  - Temporal and spatial thinking
  - Systems thinking
  - Geologic reasoning and synthesis
- Problem solving in the context of an open and dynamic system
  - Asking appropriate questions
  - Understand context of problem
  - Problem solving in 3-and 4-D
  - Ability to work on problems with no clear answers
  - Managing uncertainty in problem solving
  - Have a passion for solving problems
- Working by analogy, inference and the limits of certainty
- Intellectually flexible — applying skills in new scenarios

Technical Skills

- Problem solving with data
  - Data collection and interpretation, use of data and application
  - Evaluation of data, data quality, purpose of collecting data, begin with understanding of how data will answer question
  - Understanding data and uncertainties
  - Make predictions with limited data
  - Use of appropriate methods, reading and interpreting graphs
- Quantitative/math skills
  - Differential equations/linear algebra
  - Probability and statistics (to understand risk)
  - Understanding of scale
  - Computer programing skills (be able to think about how to solve a problem computationally)
- Experience with authentic research, collection of new information
- Critically evaluate literature, encourage critical thinking

Field and Technology Skills

- Field skills
  - Field camp and/or field mapping experiences
  - Improves spatial cognition, creative problem solving, teamwork, geoscience synthesis
  - Data supports field skills are unique and essential, difficult to replicate or substitute
- GIS — most essential for building large data sets
- Ability to handle and analyze Big Data
- Use of visual models, modeling tools (Stella, Modflow, Matlab, etc.)
- Integration of technical and quantitative skills, programming, application development
- Technological diversity (need skills and training beyond point, click, and type) — i.e., not just black box
- Preparation for life-long learning
  - How to learn and use new technology

Non-technical Skills

- Oral and written communication competency
  - Science writing and verbal communication; knowing your audience
  - Public speaking
  - Listening skills
- Project management
  - Ability to work in teams
    - Be a leader and follower
    - Don't divide work; iterative process between students with different backgrounds/disciplines
  - Goal setting
  - Conflict resolution (open minded — answer may lie in the conflict space)
  - Managing problems on the front end
  - Solution-oriented approaches
  - Time management
- Professionalism, interpersonal skills
  - Ethics, ethical awareness and conduct
  - Business acumen and risk management
  - Cultural interactions, cultural literacy, emotional literacy, learning styles, awareness of implicit bias
  - Leadership
  - Career awareness/resume/interview preparation
- Global perspective
- Understand societal relevance

RANKING OF IMPORTANCE OF SKILLS IDENTIFIED BY EMPLOYERS

Skill	Level of Mastery
Critical thinking/problem solving	Proficient
Communicate effectively with scientists & non-scientists	Proficient
Readily solve problems, especially spatial and temporal	Mastery
Make inferences about Earth system from observations of natural world combined with experimentation and modeling	Mastery
Work with uncertainty, non-uniqueness, incompleteness, ambiguity, and indirect observations	Mastery
Ability to access and integrate information from different sources and to continue to learn	Mastery
Understand and use scientific research methods	Proficient
Have strong quantitative skills and ability to apply	Proficient
Integrate data from different disciplines and apply systems thinking	Proficient
Have strong field skills and working knowledge of GIS	Mastery/ Proficient
Work in interdisciplinary teams and across cultures	Proficient
Have strong computational skills and the ability to manage and analyze large datasets	Proficient
Be technologically versatile (i.e., Google Earth®, tablets, smartphones)	Mastery

Additional Employer Discussion

- Difference between explicit versus tacit knowledge
  - To what extent to you give this versus they discover themselves
  - Transferability of skills and competencies
- Integrated culture between academia and industry
  - Advisory boards
  - Workshops and engagement
- Geoscience as a service to society
  - Discussion topic with students for recruitment and retention
  - Looking at the big picture
- Balance between geoscience knowledge and professional skills
  - Where to draw that line
    - Should there be requirements or a certification before completion?
    - Should curriculum prepare students for ASBOG exam?
  - External forces

Experiential Learning

- Vehicles for practical problem solving
  - Constant engagement in opportunities to practice skills and use concepts
  - Field experiences
  - Internships
  - Senior thesis, research projects
  - Project-based courses
- Technical skills being at the forefront
- Getting internship etc., experience — the earlier the better, the more often the better
- Use games to teach & reward innovation and creativity

Ways Employers Can Help

- Provide opportunities for end-to-end learning: conceptualization, collection of data, analysis, and presentation
- Providing samples and data
- Problem oriented short courses
- Collaborative research projects between employers and academia
- Judging student activities
- Supporting field experiences — providing funds and representatives for teaching
- Sabbatical programs — both ways; faculty fellowships

- Internships — earlier and more often the better
- Virtual internships, classes, and field experiences
- Campus teaching awards or pedagogical development
- Geoscientists without Borders® — raise profile of the science and profession training

Training

- Importance of practical applications and problem solving embedded in the curriculum
- Opportunities for instruction by professional partners
  - Professors of practice, invited experts
  - Short courses, webinars, practicums, pop-up and micro (AGI) courses in core competencies
  - Annual industry symposia, career days with emphasis on problem solving skills and essential component
  - Opportunities for partnerships with curriculum developers
- Working through industry-instructional barriers with proprietary data and software

Opportunities

- Tech transfer as a part of greater community engagement meeting the mission of both the institution and industry
- Added value of local, more frequent engagement and ties to broader impacts
- Growing/expanding stakeholders, growing campus consortiums and partnerships towards similar goals

Measuring Effectiveness and Competencies

- Accreditation of programs and practicality of course (engineering model; ASBOG)
- Collection of courses, practicums with appropriate assessments
- Need stronger articulation of measures of competence
  - Relate to specific employer needs
  - Explicitly tie skills and applications to courses
  - Acknowledging one size does not fit all

Additional Thoughts

- Models, tools, and resources already exist for how to develop the whole student and prepare them for employment
  - Cutting Edge website
  - ASBOG test as a source of problem-oriented activity for the classroom and as an incentive
  - Assessment, accreditation (demonstrable and measurable)
- Requirements for success
  - More active collaboration between academia and the outside employers
  - Ranging from formal courses to informal mentoring of students' faculty
  - Culture changes needed for faculty to place greater value on the benefits of active collaboration
  - Faculty need to be incentivized to increase experiential learning into classroom/curriculum
  - Incorporation of the non-technical skills into the geoscience curriculum or proactively include or advise students to take courses/experiences related to those non-technical skills (e.g., business, ethics, etc.)
- Solutions
  - Professional societies leading active discussions about these collaborations
  - Increased industry-academia interaction
  - NSF funding for research to demonstrate outcomes of these collaborations — we are missing opportunities

Howard Bluestine for AGI's 2014 Life in the Field Contest

Appendix C: Summary and Analysis of Departmental Progress

The 2016 Heads and Chairs Summit on the Future of Undergraduate Education had 114 participants from community colleges, bachelor's-granting four year colleges, terminal master's-granting institutions, doctoral-granting universities from R3/DPUs to R1s, plus a National Science Foundation, industry, and two professional society representatives. Before leaving the Summit, participants submitted action plans for their individual departments. Some universities had multiple attendees, including panelists and organizing committee members, who worked together on the action plans for their institution. From these Summit efforts, 79 action plans were generated.

The 2017 Earth Educators Rendezvous (EER) had 34 participants. Of these 14 had been to the 2016 Heads and Chairs Summit and had submitted an action plan progress report by the summer of 2017, and one also submitted a follow-up report in May 2019. For the heads/chairs that had not been to the previous summit, 12 submitted an action plans and 4 submitted a progress report in 2019. A second request was made in 2020. Several institutions had multiple participants and two participants were observers (NSF, SERC) and two did not submit action plans.

Overall between the Heads/Chairs Summit and EER workshop, we received 91 individual department action plans. A total of 56 participants provided feedback regarding the progress their departments had made toward reaching their Action Plan objectives (62% return), mainly between 16–18 months afterward the Summit or Workshop, however 3 from the workshop were submitted after 2 years, and 5 from the summit after 3 years. A follow-up progress report was submitted by 11 in 2019 about 2 years after the first one, and by one in 2020 after 3 years. These later and updated reports help show the overall timeline needed for making change in undergraduate programs. The responses span the range of participating institutions (5 2YC, 9 bachelor's-granting, 11 terminal-master's, 14 doctoral R2/R3/D/PU and 25 doctoral RI institutions).

To identify best practices and common problems and solutions, we asked the following questions:

- How much progress have you made with your plan? If you have modified your plan since then, in what way did it change and why?
- What has been accomplished, whether it was in your original plan or not?
- What are your future plans?
- Which implementation strategies worked — i.e., what was successful, and what wasn't?
- What were roadblocks to progress or where did problems occur? And if you were able to overcome them, what did you do?
- What did you anticipate would be a problem that wasn't?
- Any advice to others who wish to make similar changes?

Below, their progress is reviewed both objectively and subjectively to identify trends among departments.

To describe their progress objectively, their feedback was quantified using a rubric (provided at the end of appendix) designed to evaluate the common themes identified in the workshops — curriculum redesign, increase in an active learning-based pedagogy, 2YC to 4YC transfers, recruitment of underrepresented groups, adding new courses and adding/deleting majors. Curriculum redesign and increased implementation of active learning require the greatest amount of

Courtesy of the Jackson School of Geosciences, University of Texas at Austin

time and active engagement by faculty so it carried more weight when progress was being evaluated. (Not all participants proposed to address all six themes.)

Additionally, the objectives of action plans varied in detail and magnitude. Some of the smaller institutions had very specific action items such as "rework an introductory geoscience course to improve recruitment and retention" or "increase lab space and technology access" while the larger departments proposed a complete curriculum overhaul by producing backwards design matrices, planning retreats and final adoption of curricula. Finally, in terms of updating progress, some departments reported very specific accomplishments, for example they "created a Mogk-style matrix for each of three majors, identified gaps and rewrote curricula." Others reported progress in more general terms, such as "our plan has been largely implemented" without commenting on process or specific outcome.

When taking into consideration the specificity of Action Plans coupled with the variation in reporting style, the overall trend is that *the fewer the number of geology faculty in a department the more they had progressed toward implementing their plan* (Fig. C-1). When 'Carnegie classification' is considered, 2YC, BA/BS and MA/MS institutions made more progress than R1 and R2 institutions, even when R1s and R2s had equivalently small faculty sizes.

In some departments, the slow rate of progress was attributed to common factors such as lack of faculty buy-in and bureaucracy (curriculum approval, funding for new faculty, lab space, equipment, etc.). Eleven departments had begun a campus or state-mandated curriculum overhaul before the first Summit or workshop, so they were already a few years into the process when they attended. As is evident from the chart, where faculty size exceeded 27, departments did not make more than 50% progress. Although faculty numbers between 5 and 10 appear to be most successful, too few faculty was a challenge as they were spread too thin to pursue a deep dive into curriculum redesign and to engage in new teaching methods. Their focus was primarily on recruitment/retention

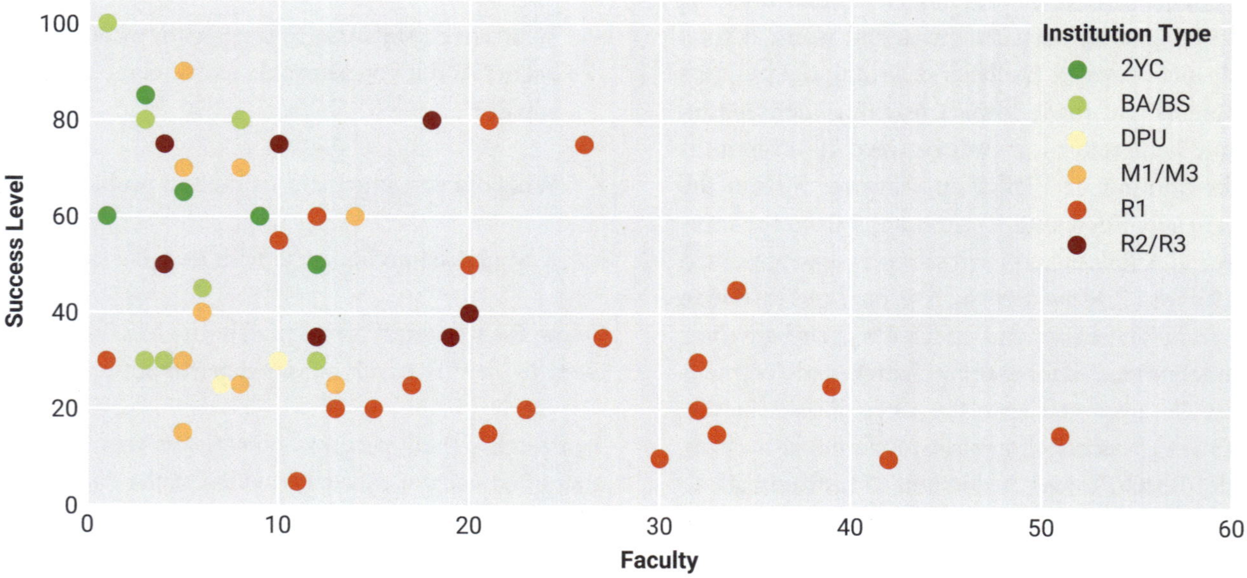

Figure C-1: Reported Program Reform Success

Relationship by faculty size and instition type

Reform success as related to faculty size and institution type.

and 2YC to 4YC transition. Had departments had more time or if they had already begun curriculum revisions, they likely have made more progress.

Twelve departments submitted a second progress report and all but two reflected more progress, primarily because they had more time to achieve results. Common themes from those who provided 2019 updates to their 2017 progress reports:

- change is a continuous process that requires patience
- the ones who sent updates generally had a defined vision and stewarded progress against that vision
- short-term goals and tactics were continually updated in response to internal/external changes — university mandates, staff changes, etc.
- over the 2-yr period, departments had commonly focused on one aspect of their plan
- continuity of champions was an issue for those who didn't make much progress
- Five Ph.D. granting institutions are good examples of developing vision, empowering others to act on vision, addressing obstacles, and finding short-term wins

Carnegie Classification	Faculty Size	Progress from Report 1 to 2
Bachelor's (Public)	4	30/30
Bachelor's (Public)	12	30/65
M1 (Public)	5	35/75
M1 (Public, MSI/HIS)	5	15/15
D/PU (Public)	10	30/50
R2 (Public)	19	35/60
R2 (Public, MSI/HIS)	20	40/65
R3 (Public)	10	75/85
R1 (Public)	11	5/10
R1 (Public)	20	50/70
R1 (Public)	27	35/85
R1 (Public)	42	10/45

Table C-1. Progress by institution type and faculty size

From a more subjective perspective, many of the departments shared similar experiences in terms of successes and challenges. Collegiality, support of leadership, department retreats and the use of SERC-endorsed resources contributed to successes. Many were constrained by outside forces such as limited campus budgets, hiring freezes and low enrollment. No matter the department size, many departments had to contend with faculty who were unwilling to change traditional curriculum

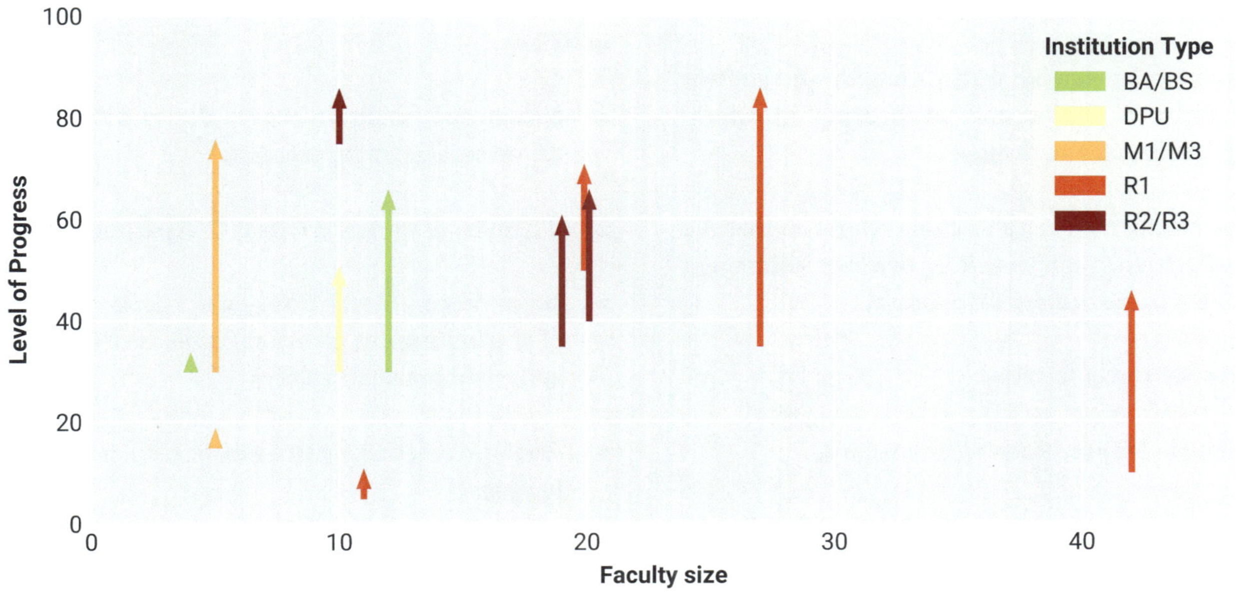

Figure C-2: Reported Progress on Reform

Departments providing additional updated progress reports

Reported progress on reform as reported by departments.

or adopt new pedagogy, but with persistence many succeeded. More details are provided in a later section.

ELEMENTS OF SUCCESS

Figure C-1 clearly shows that for faculty size greater than 20 (all R1s) and for those less than 20 (all types) some were more successful than others, leading to the following questions:

For the four R1s with 20 or more faculty that were over or at 60% successful — what did they do that was different? Why were they successful? What part did they primarily distinguish themselves doing — i.e., was it redoing the curriculum, changing pedagogy, etc.

R1, >20 faculty, > 60% successful: Two of four departments submitted two progress reports (one improved from 40% to 60% and the other from 35% to 85%):

All four redesigned their curriculum using the backwards design matrix approach. Only one of the four substantially addressed pedagogy. Two provided dedicated time for a faculty member to work on program changes.

Additionally, departments noted:

- increased a faculty member with a 9-month appointment to 12-months for the purpose of focusing on career-building exercises for students
- granted a faculty member a teaching release so that she could dedicate attention to writing new curriculum and submitting it to the campus for approval
- involved early-career faculty
- implemented new peer teaching evaluations
- implemented training course for grad students to improve TA support of new active learning courses
- considerable sharing of course syllabi among faculty to minimize overlap in prerequisites and core courses
- produced fully developed matrices at a retreat devoted to this purpose and attended by all faculty
- many major courses were flipped
- assessing outcomes (to show success!) via campus mandated degree assessment

Same questions for the three R1s with more than 20 faculty that were between 40–50%. And for the 3 R2s with close to 20 faculty and over 60% successful?

R1, >20 faculty, 40–50% success (3 departments, 1 of 3 provided two progress reports but did not make progress between the two):

- curriculum effort was driven by a few faculty with a bottom-up approach rather than top-down
- moved very slowly (took four years) and took time to get feedback from all faculty to assure support
- in terms of pedagogy, one department supported professional development for teaching

R2, ~20 faculty, > 60% success (3 departments; 2 of 3 provided two progress reports improving from 35 to 60% and 40 to 65%):

- encouraged faculty to engage classes with more active/experiential learning exercises
- did much of work at a retreat (2 departments)
- began process of restructuring the department with goal of sharing courses across curricula (geology, geography and environmental science)
- Backwards design matrix was instrumental in identifying gaps
- pairing courses throughout the undergrad trajectory for students to see connections and afford opportunity to work on projects as teams

- streamline programs/courses to assure they reflect what the department has the expertise to teach

For all with smaller faculty (under 20) and over 60% successful, was the success primarily because of major curriculum or pedagogy changes or was it all the other types of things, like e-portfolios or increasing some skills/concepts like more communication, etc.

Of the 19 departments:

- One-third of them had already begun curriculum redesign prior to attending workshops which boosted their progress (6 departments)
- More than half made changes to pedagogy (11 departments)
- Less than half tackled other areas such as 2YC–4YC transfers (7 departments), improving recruitment and retention (6), adding new courses (4), incorporating e-portfolios (2) and making changes to majors (4). Most of the departments that tackled these other areas fall in the 5 to 10 faculty range. (*See below for specific departments)
- Three of the four 2-Year colleges did not propose to make curricular changes at all and while they may have done some tweaking, their progress was based on improvements focused primarily on recruitment/retention and transfer to 4-year programs
- Hosted an NAGT Traveling Workshop, "Building Stronger Geoscience Departments"

***Faculty <20 faculty and >60% success, areas outside of curriculum and pedagogy:** Most of the departments that tackled these areas fall in the 5 to 10 faculty size range.

2YC–4YC transfers:
7 departments — 2 2YC, 2 BS/BA, 2 M1, 1 R3/D/PU

Improved recruitment and retention:
6 departments — 3 BS/BA, 3 M1

Added new courses:
4 departments — 1 BS/BA, 1 M3, 1 M1, 1 R2

Incorporated e-portfolios:
2 departments — 1 BS/BA, 1 R3/D/PU

Made changes to majors:
4 departments — 2 BS/BA, 1 M3, 1 M1

Are there any strategies that clearly were a major failure — not including budget cuts, upper administration issues, etc.?

No, no major failures. Instead of sweeping curricular or pedagogical improvements, one R1 department put a lot of energy into a proposal aimed at improving relations with local two-year colleges that did not get funded. It did, however, build a relationship with the science education department.

Successful Strategies

An additional analysis based on a set of questions provided the following information:

1. How useful were retreats?

Extremely. Not one reported failure. Nine departments planned retreats and seven had executed before their last progress report:

- Bachelor's (1 department)
  - *one planned.*
- Master's M1/M3 (2 departments)
  - Most helpful for curricular revision was two-day workshop — developed the very specific action plan with faculty names and deadlines attached for each task. *One planned.*
- Doctoral/Professional (1 department)
  - Two-day retreat with facilitators from NAGT's Traveling Workshop Program was very much a success.
- Doctoral R2 (3 departments)
  - Where the most progress was made.
- Doctoral R1 (3 departments)
  - SERC pages on retreat planning very helpful.

- All-day retreat was REALLY worthwhile — did it all the way across campus with meals catered. That bought eight hours x 40 faculty required to get the basic data, which a smaller group of staff and leadership could then use to create the final matrices.
- Successful Friday afternoon faculty retreat to discuss curriculum map. *One planned.*
- We abandoned the Mogk matrix when it became clear that it was going to bog us down in an overly detailed process at the expense of the big-picture reforms our curriculum needed. This decision to approach the reform in a phased way, with big-picture reform first, and course-by-course matrix analysis later, is what saved our process.

2. How useful was doing the backwards design/matrix?

Fifteen of forty-eight departments that addressed curriculum specifically referenced the matrix approach or backward design.

- Master's M1/M3 (4 departments)
  - One department brought in David Mogk to present a seminar and afternoon workshop on curriculum planning and content.
- Doctoral R2/R3/D/PU (6 departments)
  - Backward design or matrix used.
  - Constructed a matrix of undergraduate course competencies/skills. Posted the matrix online and faculty identified which are taught in their courses and which should be added to undergraduate curriculum.
  - Matrix identified major gaps (however 50% of faculty either misinterpreted or neglected request for input).
- Doctoral R1 (8 departments)
  - We collected syllabi and materials and met individually with all faculty to verify which outcomes and skills were met by the course and at what level (basic, enriched, reinforced). We have assessed our outcomes (one course per year) via mandated degree assessment from the University starting with GEOL 101 and moving up through the curriculum.
  - Matrix identified gaps in quantitative skills and math.
  - Major curriculum revision in 2014 lacked a formal document to identify which concepts, skills, and competencies are covered in the new curriculum so retroactively developed such a metric.

3. What kinds of things made changing to more active learning pedagogy easier?

In general, attending workshops (SAGE, Earth Educators' Rendezvous); open faculty discussion and sharing of resources; enthusiastic junior faculty; positive outcomes supported by student assessment; and introduction of new ideas at retreats.

More specifically:

- 2 Year Colleges (4 departments)
  - Explore topics through investigation using physical samples and models plus technology using large data sets, interactivity, animations and video instruction.
  - Successfully transitioned to a flipped classroom model for introductory course which focuses on active learning vs. passive listening.
- Bachelor's (1 department)
- Master's M1/M3 (2 departments)
  - Progress establishing the importance of successful adoption of active-learning in courses as a faculty performance expectation.
  - Two geoscience faculty were instrumental in establishing a college FAST Team (Faculty Advocates for STEM Transformation) that is working to showcase active-learning pedagogies and to provide professional development.
- Doctoral R2/R3/DPU (3 departments)
  - Incentivized our merit review process to reward faculty who develop and implement reformed teaching methods that incorporate engaged learning.

- Doctoral R1 (4 departments)
  - Peer mentoring in active-learning introductory classrooms. We have "new" faculty co-teach these courses with seasoned instructors and then take over the course as lead instructor after 1–2 semesters. With help of teaching post doc introductory lab course was "flipped" and Mineralogy, Geochemistry and Biogeochemistry are all major courses that have been "flipped".
  - Identified intro level courses taken by all majors that make them ideal candidates for innovative instructional methods such as the flipped classroom and blended learning. Seeking new faculty hires who have experience in innovative instructional methods, and encourage new faculty to bring new and creative methods to our department.
  - 3 faculty have been supported to undertake professional development for teaching and 3 more have been supported in developing experiential international courses. Chair attended a SAGE workshop where some pedagogical methods were summarized and demonstrated.
  - Recognizing that classic or traditional freshmen geology classes are becoming less attractive, we were motivated to be proactive in what we teach and had the willingness to do our best to offer a modern geology courses in our curriculum.

4. Were there specific comments about using the employer's recommendations for concepts and skills or the Summit outcomes being used to demonstrate what needed to change (either course content/courses or incorporation of skills)?

Yes, some examples:

- Bachelor's
  - All elective courses focus on developing skills and conceptual understandings valued by industry, including quantitative, written, conceptual and technical skills.

- Master's M1/M3
  - Sharing summaries from the Summits, along with the employer-vetted concept and skills matrix, helped greatly with faculty buy-in.

- Doctoral R2/R3/D/PU
  - Focused on streamlining student learning outcomes and methods of assessment by integrating more critical thinking, writing, presentation skills, and teamwork into existing classes. There has also been a concerted effort to encourage faculty to engage classes with more active/experiential learning exercises.
  - Curriculum that stresses repeated exposure and expected mastery of key skills necessary for conducting research, reporting research (oral and written), analyzing data and designing research plans.

- Doctoral R1
  - Emphasized topics such as experiential learning and field instruction rather than specific content.
  - Discovered had placed much of the transferrable skills in specific classes (e.g., writing, in one course titled Scientific Communication) which meant students would get these skills once, but maybe not again. Alpha-testing infusing these transferrable skills into the majors courses (e.g., writing exercises, taught by a second instructor who specializes in scientific writing, embedded in the context of an exercise in Structural Geology).

5. Any mention of addressing specific skills (quantitative, communication, field, etc.) in their curriculum?

Yes, in addition to the previous question:

- 2 Year Colleges
  - Purchase of E-science lab kits geared toward testing and measuring/analyzing hypotheses.

- Bachelor's
  - Intentionally embedding science communication into key parts of the curriculum, helping students to increase their capacity for effective written, oral, and graphic communication to both scientific and non-specialist audiences.
  - Produced a 'Geology Roadmap' given to all majors that outlines discipline-specific skills and student learning outcomes.

- Although a set of content-specific courses, new program represents a collective, conscious shift of focus toward building core student competencies and "must-have" skill sets in addition to content-driven knowledge base. Placing an even greater emphasis on students using, applying, communicating and being innovative with what they've learned.
- To expose students to research invoke more research-like experiences earlier in the curriculum to bring all students to a higher level of readiness.
- Integrating more critical thinking, writing, presentation skills, and teamwork into existing classes.

▶ Doctoral/Professional
- Developed an Earth Science Professional Development Seminar course for all Earth Science majors. The professional development course helps students become aware of career prep resources on campus and in the discipline, learn to write resumes, practice interviews via video conferencing, etc.

6. How many worked on 2YC to 4YC transfers—how many had success?

10 departments made progress, from building relationships with institutions to the following:

▶ 2 Year Colleges (3 departments)
- inviting 4YC faculty to field trips for one-on-one time with students; student panels and socials for discussing transfer options.

▶ Master's M1/M3 (2 departments)
- 13 new transfers into 4YC in one summer.
- Hosted a conference for state system campuses and was well-attended by 2YCs.

▶ Doctoral R2/R3/DPU (3 departments)
- Multiple summits with area 2YC instructors.
- Submitted NSF Pathways grant for connection with 2YC (not funded).
- Completed articulation agreements for the Geology degree with 4 local two-year community colleges and are nearly done with a fifth agreement. We have continued to work on strengthening our connections with those two-year colleges.

▶ Doctoral R1 (2 departments)
- Developed a transfer pathway/academic map for students transferring in from 2YC. Developed freshman field exchange with local CC geology faculty.

7. How many reported on recruitment and retention of underrepresented minorities or recruitment/retention overall? Any specific examples of success?

8 reported on recruitment and retention. Some success stories:

▶ 2 Year College (1 department)
- Faculty attendance at campus outreach events for international students and veterans.

▶ Bachelor's (4 departments)
- To increase retention of first gen college students, designed a 1-credit course that includes problem-based field and lab activities, discussions, and visits from alumni and industry-professionals. Students will leave with a clear path to graduation, an outline for a portfolio and resume, and knowledge of various career opportunities.
- Focus on messaging with Admissions office recruiting and advertising materials; inclusive of disparate levels of science background.
- 50% faculty now from underrepresented groups; active learning helps retention.

▶ Master's M1 (1 department)
- Student mentoring/support for diverse students; collaboration with other STEM departments

▶ Doctoral R2/R3/DPU (2 departments)
- Presentations during introductory classes informing students about the Geology major and employment opportunities
- A key part of the IUSE GEOPATHS grant, the undergraduate research cohorts, has been successful with a substantial increase in the number of B.S. graduates going on for graduate work, including 3 students from underrepresented groups (African-American and Latina). All of those who went on for graduate work were connected in some way to the research cohorts.

▶ Doctoral R1 (1 department)

8. How many reported on K–12 teacher training, how many successes, what type of institutions did this (i.e., # of each)?

Only two

- 2 Year Colleges (1 department)
  - Admin focus

- Master's MA/S (1 department)
  - Hoped to link matrices to Next Gen Science Standards for benefit of science education students but received huge faculty blowback (currently working on link to concepts/competencies/skills as well as ASBOG Fundamentals instead).

9. How many talked about working with employers? Note — not necessarily in many action plans.

Very few.

- 2 Year Colleges (2 departments)
  - Worked with local museum and city government for internships.
  - Local community organizations have created undergraduate research opportunities for students in a service learning capacity.

- Bachelor's (1 department)
  - Bringing in industry representatives to discuss career paths.

- Master's M1/M3 (3 departments)
  - Representatives from the State Board (that employs many geoscientists) spoke to faculty and students; this was effective in presenting competencies vetted by employers.
  - County Water Dept. for paid internships and fieldwork.
  - We have shared the matrix with our advisory board and they have provided feedback related to the workforce skills need in their particular segment of the geoscience workforce.

- Doctoral R2/R3/D/PU (1 department)
  - The "Career Track" will prepare students for graduate school or employment in the field of geosciences. The "General Track" provides the student with a solid background in the geosciences, however it is probably lacking in some of the cognate courses (specifically Calculus, and less rigorous Chemistry and Physics). Second track was developed for students who enjoy geology but may end up working outside of the geosciences.

- Doctoral R1 (1 department)
  - Our Advisory Council (consisting of employers) reviewed and discussed the Summit series outcomes and were in general agreement with the results. We continually work with employers of our students with them providing students career and interviewing advice, giving professional talks and providing datasets.

10. How many used e-portfolios?

Five total:

- Bachelor's (1 department)
  - Students in a single credit 'career path' class outline an ePortfolio

- Doctoral R2/R3/D/PU (4 departments)
  - Department developing an ePortfolio template to illustrate mastery of each competency/skill which will also include a student reflection component.
  - Used ePortfolio assignments to assess student achievement of more specific geoscience knowledge and skills, such as those identified by the Summit participants.
  - AY2019–2020 implemented requirement that all students in specific Geoscience courses complete a 'signature assignment' to be uploaded to an ePortfolio. At the end of each semester, student ePortfolio 'signature assignments' are assessed on several specific criteria. A university-provided rubric was used as a rating tool for ePortfolios. A small financial incentive (paid by the university) to faculty willing to implement the ePortfolio requirement in their courses.
  - May add an "ePortfolio" based on competencies as a capstone requirement.

11. Any comments on assessment strategies?

Yes, 15 include the need for assessment in their plans but only a handful offer successful strategies:

- Bachelor's (3 departments)
  - Challenged with annual assessment exercises because although learn what graduates know, it's not clear where in curriculum they learned it
  - Faculty have been asked to assess WOVN — Writing, Oral, Verbal, Numerical — in every course. Also, add these questions to annual performance review: What do you do? What have you tried? How do you know that changes you make are better? Are you still teaching what students can look up? Chair included the first three questions in performance and course reviews AY 2016–17. The last question will be added to course and performance reviews in AY2017–18 in coordination with better implementation of introducing active learning methods.
- Master's M1/M3 (3 departments)
  - Generating questions for an exit interview for graduates.
- Doctoral R2/R3/D/PU (3 departments)
  - Considering replacement of University-provided rubric used for ePortfolio assignments to assess student achievement with more specific criteria identified by the Summit or combine the two.
  - Assessment resulted from merging of our original plan to use ePortfolio assignments to assess student achievement of more specific geoscience knowledge and skills, such as those identified by the Summit participants, with the University's new Quality Enhancement Plan (QEP).
- Doctoral R1 (6 departments)
  - Looking at alternative assessment (pre-post- test in all courses — bridging between previous and next course in the sequence).
  - Developing a database of what graduates do in terms of jobs, further education, and careers but challenge to obtain necessary contact information from the administration.

PROGRESS REPORTS SUMMARY

Six major themes were identified in the progress reports: changes focused on:

1. curriculum redesign
2. pedagogy
3. promoting 2YC–4YC transfer
4. recruitment and retention of underrepresented groups
5. changes to the major,
6. changes (addition/deletion) of courses.

Progress reports are discussed according to institution type (2 Year Colleges, Bachelor's public, Bachelor's private, terminal Master's granting institutions, Doctoral DPU, R3, R2, and R1. The following summarizes the experiences of the different departments in implementing their action plans, which strategies promoted progress, what were the challenges and roadblocks that impeded progress, and any advice.

Five 2YC institutions range in faculty size from 1–5:

A total of 5 two-year colleges submitted progress reports in 2017 and another one in 2019. Of these, one institution has plans to implement a **curriculum redesign** assisted by a departmental matrix. With respect to changes in **pedagogy**, 4 of these institutions have made progress through the implementation of service learning, flipped classrooms, open-access educational resources, and competency-based review. In addition, interest in pedagogical changes have resulted in increased participation in faculty professional development. One of the 2 Year Colleges mentioned increased **collaboration with four-year colleges**, which resulted in new learning outcomes and partnership with environmental (rather than just geology or earth science) programs. The 2019 report mentioned importance of collaboration with the community for internships, other student opportunities.

Successes and Strategies

Developing relationships is critical for transitioning students to 4YC. Invited faculty members from possible transfer universities to campus or on field trips for students to learn about future opportunities.

Connecting our students with internship opportunities is important. A grant supported students from a community college to work the summer with UNAVCO in Boulder.

Collaborating with the community, including a local museum and city government, increases outreach and education beyond community college's walls. This, of course, improves recruitment and retention of students.

Challenges and Roadblocks

A large part of the focus is on teacher education and training, which is a positive outcome but can deter efforts to produce successful geoscientists (curriculum design, learning experiences).

University-wide budget issues caused a hiring freeze.

Time is the greatest barrier — a lot of these changes require time to write, propose, collaborate, etc... other duties have to take priority (2 departments).

Advice

Just hang in there — follow highs and lows of budgets and dynamics of leadership with their own goals for campus success.

Developing relationships is critical. We had a weekend field trip and invited a faculty member from the transfer university to attend. Having that unstructured time for majors to ask her questions was invaluable.

Obviously "be the change." You have to model the change for others to see what works. And don't be afraid to ask what others are doing in their classrooms. Encourage them to make student-friendly decisions.

Anna Stanczyk for AGI's 2017 Life as a Geoscientist contest

Six BA/BS institutions range in faculty size from 3–12:

BA/BS institutions that submitted progress reports include 6 institutions, with between 12 and 4 faculty (with an average of 6 faculty). Two of the institutions sent 2019 updated progress reports focusing on the need to evaluate how effectively the changes are serving students. Of these institutions, 5 worked on **curriculum redesign** — Electives were modified to add competencies, and some courses were eliminated. Learning outcomes were added; science communication was emphasized; a roadmap for majors was created; more common core courses were required, and a shift to competencies and skills occurred. Other institutions still have plans to reconfigure their curriculum. Major advice regarding curriculum changes include: work on big picture engagement and agreement through communication and dialogue, and course level change is difficult, even with big picture agreement. One of the major strategies was to allow the younger faculty take lead.

Four institutions made progress on **pedagogy** changes, with the introduction of more active learning techniques and more focus on measurable outcomes and assessment. Faculty discussions, departmental retreats, and increased participation in teaching workshops were instrumental in facilitating these pedagogical changes.

None of the BA/BS institutions worked on **2YC–4YC transfers.** Four of these institutions worked on promoting inclusion of **underrepresented groups.** A new seminar course was designed (but not yet implemented) to aid in retention. General outreach efforts, as well as specific effort to make the department more accessible for students with varying levels of science readiness, were enacted to work on recruitment. Strategies included faculty involvement directly in the retention and outreach efforts, as well as active learning as a means of promoting retention.

Two institutions worked on **changes to majors**, introducing a cross-disciplinary BS in geobusiness, a new environmental geology degree, a new geophysics major, a non-thesis MS in Environmental Science, and new certificates. These new initiatives were helped by focusing on cross-disciplinary degrees, and through work with faculty and directly with the academic vice president. One institution worked on changes to **courses**, adding a statistics and a GIS course.

✯ Successes and Strategies

Faculty receptive to programmatic changes.

Overall faculty engagement. — i.e., understanding the difference between course assessment and student assessment.

Defining the big picture in terms of incorporating a competency based approach.

Established a new Geology program — although on paper it is a set of content-specific courses, it represents a collective, conscious shift of focus toward building core student competencies and "must-have" skill sets and attributes in addition to content-driven knowledge base. Now placing greater emphasis on students using, applying, communicating and being innovative with what they've learned. The new program design now fills more gaps and contains more purposeful scaffolding.

We led an effort with the business and math departments to propose a new cross-disciplinary B.S. degree in GeoBusiness and Data Analytics. Students take core geology classes and courses in business, economics, statistics, big data management, spatial data analysis, programming, and technical writing. The integrated skill set developed was viewed by our advisory board and other industry experts as very strong and employable. The Academic Vice President sees it as an innovative degree that will prepare students for the business workforce and for non-geology graduate work (MBA, geo-economics, etc.); he is encouraging other departments to consider similar cross-disciplinary degrees.

Many of our students have an interest in geospatial studies so we created a B.S. degree that builds on that interest that includes core geology classes as well as earning a certificate in GIS and a certificate in another technical field including computer programming, web design, networking, or data science. Graduates with this degree have found immediate employment.

Leveraged annual program review and mapping core competencies onto departmental and institutional learning outcomes and constructing a competencies matrix to make sure that students acquire core competencies and desired skills through our program.

Existing curricular map was evaluated. As a department we agreed that all elective courses should focus on developing skills and conceptual understandings valued by industry, including quantitative, written, conceptual and technical skills.

Student assessment will be added to course and performance reviews in AY2017–18.

Our network with internship-granting agencies has grown and more students are taking internship for credit. Also, some members of our faculty are part of a state-level team attempting to launch a state system-wide geology field course for Geology students at the 14 universities.

In response to historically low graduation rates, we designed a 1-credit *'active learning' class* during which students design a clear path to graduation, an outline for a portfolio and resume, and are exposed to various career opportunities.

To meet the needs of students of varied interests and backgrounds, we developed an Environmental Geology degree and worked with a nearby university to create a *non-thesis 4+1 M.S. option in Environmental Science.* Early measures of this degree suggest about 70% of graduates will go, or have gone directly into industry, and about 20% are interested in graduate school.

University-mandated reduction of classes with low enrollment resulted in reducing elective courses from nine to six. The remaining six electives strengthened their focus on developing skills and conceptual understanding valued by industry. (Direction from above contributed to the outcome.)

△ Challenges and Roadblocks

While colleagues were very receptive to programmatic "big picture" changes, affecting what they do in the classroom is much more difficult.

Quantitative assessment is tricky. Often it seems as if a new active-learning activity is ephemeral, the students feel good about it, they enjoy it, but there is little to no improvement in the overall learning in the class.

Implementation efforts were derailed by bureaucratic forces and the limits of smaller faculty size. Of the three full time faculty, one retired and the college will not authorize a replacement which leaves the remaining two to cover all department duties.

State budget cuts and declining enrollment preclude funding for professional development even though senior administration is aware of the need to incentivize teaching.

Biggest obstacle remains a budget shortfall and a decade of plummeting enrollment.

Progress has stalled because of major budget constraints due to declining enrollment.

Advice

Open dialogue and communication within the department is key.

Leverage institutional processes, such as regular Program Review, to drive change.

Add these questions about teaching to annual performance review: What do you do? What have you tried? How do you know that changes you make are better? Are you still teaching what students can look up?

The nine M1 and two M3 institutions range in faculty size from 4–14:

Seven MS/MA institutions submitted 2017 progress reports. One of those institutions one submitted a minor 2019 update, and 2 institutions submitted new reports, for a total of 9 MS/MA institutions reporting. Eight of the nine institutions made progress on **curriculum redesign.** Progress included revisions to courses and curriculum, related to an NSF-IUSE project; increase resources to add engineering geology and geophysics programs; and development of a curriculum matrix combined with internally developed Knowledge, Skills, and Dispositions work. Additional changes include design of upper-level courses to fill gaps, improved internship and undergraduate research opportunities, collaboration with a civil engineering school to design a 3–2 curriculum to recruit students with stronger math backgrounds; and addition of physics, math, GIS, surface processes and field components. Future efforts include a department retreat to help guide curriculum change, new faculty hires; and updates to BSc with more focus on workforce skills. Progress was possible with faculty who are motivated to make changes, collaboration with other departments, and capitalization with internal experiences and NAGT site visits.

Progress related to **pedagogy** includes faculty mentoring and peer evaluation, adoption of active learning as a faculty performance expectation. Other progress includes: newer faculty are working with a geology lecturer to build reformed (active) teaching modules. College-wide effort to increase active learning in STEM departments. Faculty developed team to work on showcasing these pedagogies and providing professional development.

Two institutions are working on **2YC–4YC transfer**, with 2YCs visiting to develop a pathways document and to attend a meeting on the community consensus vision. One institution reports that these efforts have led to noticeable improvement in attracting 2YC students to program. 2YCs have been involved in conversations regarding skills, competencies. These MA/MS institutions advise to build strong relationship with 2YC faculty (faculty–faculty > administrator–administrator).

One master's institution has made progress on **underrepresented groups**, with a cohort-based model to recruit and retain students in all STEM departments, and training for faculty on improving curriculum and student support specifically for "more diverse" students. These initiatives have been designed and will be implemented in the future. This initiative was successful due to collaboration with other departments (biology) to secure a HHMI grant.

Three institutions have instituted some **change to the major**, with a two-phase declaration system, and new certificates in geospatial technologies. There are also plans to provide Engineering Geology programs, an MS in Geophysics, certificate in climate, and major in climatology. **Changes to courses** were made at two institutions, with upper level courses being redesigned to fill curriculum gaps, addition of a Tectonics course, addition of Hydrogeology as a core course, and a new general geology course in Natural Hazards. Close advising by faculty has encouraged students to enroll in elective courses. Enrollment in a general education course was disappointing.

Other progress from these institutions included a revised plan for faculty hires and successfully receiving requested new faculty line. Work with county water department on new groundwater monitoring wells provides new internships, teaching opportunities, data for research, recruitment, outreach to K–12. An effort to offer Geologic Hazards and Resources as dual credit with high school was ultimately cancelled by the high school.

✯ Successes and Strategies

Hosted an *NAGT Traveling Workshop* called "Building Stronger Geoscience Departments" where created an action plan for curricular revision and followed through with numerous department meetings dedicated to developing a new curricular model, including sharing it with a student focus group for feedback.

Open communication and complete transparency with all stakeholders. Sharing summaries from the Summits, along with the employer-vetted concept and skills matrix, helped greatly with faculty "buy in."

Spreading the word about the collective community agreement on what undergraduates should learn and be able to do from the Summit initiative to colleagues across the system.

Backwards design was very helpful—starting by identifying key student skills and knowledge, then defining student learning outcomes and then mapping these to a course matrix. Perhaps the most helpful was the two-day workshop during which we developed the very specific action plan (with faculty names and deadlines attached for each task) for curricular revision.

Made a matrix that is a progression of courses and our previously articulated Knowledge Skills and Dispositions to map progression (not just coverage), then inserted our assessment outcomes into the matrix to inform curricular discussion.

Best strategy is to allow faculty to take ownership of process—slows work, but best way forward.

Personal advising by Geology faculty, coupled with curricular changes increasing the number of compulsory classes at the expense of electives worked best.

Collaborate, and be involved positively with your college. We (Geology) became a model for other departments to use to develop assessment plans university-wide. On accreditation site visit the vice-provost in charge of managing the assessment program and developing the materials for the university's accreditation, cited our models as being critical to the university's successful accreditation review.

As an example of unintended consequences, in an effort to recruit first year students, we collaborated with other chairs in Biology, Chemistry, and Science Education to develop and submit a successful proposal to the Howard Hughes Medical Institute (HHMI). This project presented plans to develop a cohort-based model to recruit and retain diverse students into STEM departments (in this case, Biology, Chemistry, Geology, and Physics) and institute training and mentoring for faculty to improve our curriculum and student mentoring/support for more diverse students.

External facilitators helped work through the initial stages of curricular reform. It was valuable for all to dedicate uninterrupted time to thinking deeply about curricular goals, what is being done, what to do, and to be able to discuss these ideas and concerns freely among the group.

Invited representatives from the State Board (employer of students). They were very receptive to the request and their presentations were highly motivating to faculty and students. An outside authority represents competencies vetted by employers.

△ Challenges and Roadblocks

Faculty resistance to change—but two new faculty members (out of 6) and their energy and commitment to this process are really helping us move forward.

Personalities has been the biggest one. It has been more important to "win" than to actually develop functional programs.

Faculty took different approaches on course survey for skills and concepts and the reliability of the data is being questioned by faculty as a whole. I wish I would have been more organized in the way we collect this initial data.

Time.

New university mandates, strategic plans and initiatives have taken up all our time.

Limited time available for the few faculty (small department) to execute plans.

Workload for faculty heavy.

Significant decrease in number of faculty. Lost a faculty line due to low enrollments.

Smaller departments are scrambling to stay above water with limited people and resources.

Sidetracked by our campus converting from a quarter to a semester system and creation of a new general education program.

Overcame obstacles because I became Associate Provost at the university.

Associate dean/dean thought BS geology program had too many credits required and was too rigorous — scaled back slightly without giving up on rigor that we think is important if our graduates are to be competitive.

Upper Administration: Main roadblock is our dean who does not work well with departments in the college.

Lack of support from University Administration, which continues to impose its own set of changes on departments without consulting the faculty and is putting pressure on to increase the number of majors in Geology.

Advice

Use the concepts and skills matrix to your advantage, as an instrument that was nationally vetted by geoscience faculty and employers.

Key step for us was bringing in external facilitators (NAGT traveling workshop) to help us work through the initial stages of curricular reform. It was valuable for all of us to dedicate uninterrupted time to thinking deeply about our curricular goals, what we do, what we want to do, and to be able to discuss these ideas and concerns freely among the group.

Planning is everything, but actual implementation takes time. Planning should include personal networking and faculty buy-in internally (departmental level) and externally (Senate level). The departmental plan should link to the university's strategic plan. The plan should be based on data, references, and examples; the plan should include SWOT analyses and link to program review (self-study).

Be persistent, encouraging, and don't expect things to change overnight.

Slow down the pace of changes, as faculty, students, and administrators tend to resist changes/ new ideas.

Take the long view; this process takes significant time.

Congenial working relationships are key.

Collaborate, and be involved positively with your college. NAGT site visit (the Building Strong Geoscience Departments program) helped us develop and implement a good course and curriculum assessment plan. This turned out to be important, as we (Geology) became a model for other departments to use to develop assessment plans university-wide. This positive feedback has been affirming to our faculty, and will prove (I hope) useful in resource decisions in the future.

Invite representatives from your State Board (or other agency that hires students). We found them to be very receptive to our request and their presentations were highly motivating to faculty and students.

The six R2 and four R3/DPU institutions range in faculty size from 4–20:

Four doctoral/professional universities submitted progress reports between 2017 and 2020, and two submitted a 2019 or 2020 update. In the field of **curriculum**, one institution made progress in the use of the competencies and skills matrix, and faculty identified skills/competencies to add to the curriculum.

Another revised their Environmental Science curriculum to include classes with more rigor and with a broader range across the Earth Sciences and developed two new required courses for all geoscience majors — a discipline-oriented statistical analysis course and an Earth Science Professional Development Seminar course. Plans were made to add e-portfolios and to review general education program in geology and physical geography. Two institutions addressed **pedagogy.** Geology courses were included in a campus wide initiative to increase student awareness of integrated content across courses. A competency-based approach was implemented. The other initiated discussion of active learning strategies and course content in faculty meetings. Two institutions addressed **2YC–4YC** transfer. Meetings between 4YC and 2YC colleges resulted in the development of a transfer pathway/academic map, and approved by Board of Regents. Articulation agreements were completed with four 2YCs and connections were strengthened. One institution addressed **underrepresented populations**, by discussing presentations for introductory classes on the major and career paths. One developed research cohorts that resulted in an increase in the number of minorities going onto graduate school. Future plans to pursue direct contact with promising students, and 2YC students. One institution implemented **changes to the major**, developing a "career-track" and a "general-track", increasing flexibility for electives and specialization, and numbers of majors (especially transfer from Engineering and Physics) is increasing. One institution has plans for **changing courses**, and hiring a new assistant professor with computational geosciences experience to teach courses in technical data skills. An institution that provided a 2019 update reported progress revising Geography curriculum, and the one that provided a 2020 update reported that they developed an introductory-level course called 'Our Violent Earth' which includes the most exciting elements of geology and meteorology in hopes of getting a broader range of students interested in the geosciences. One started an alumni listserve.

Four doctoral R2 universities submitted 2017 progress reports, 2 provided 2019 updates, and 2 additional institutions provided new 2019 progress reports, for a total of 6 institutions reporting. Five of these institutions addressed **curriculum redesign,** identifying gaps using the backwards design matrix and having conversations on learning outcomes. Formal curriculum plans, developing unifying themes and lasting knowledge and skills across degree programs is still a goal. One institution developed new BA degree in Environmental Sustainability. Two institutions addressed **pedagogy**, disseminating literature on best teaching practices, incentivizing merit review process to reward faculty who develop and implement reformed teaching methods and streamlining assessment methods. There are plans to redesign introductory geology lectures to include more engaged learning.

One institution met at the summit with local area **2YC instructors**, revising transfer plans and disseminating them to college recruitment offices. No institutions addressed **underrepresented populations** or **changes to majors**. Two institutions made **changes to courses**, addressing issues of uneven course content across sections and adding engaged learning laboratories. Additional progress at doctoral R2 universities included implementation of differential teaching/research/service so faculty with varying roles can be credited for effort. They created a 5-year strategic plan to prioritize new faculty and staff hires, explored and rejected idea to create School of Earth and Environmental science, and implement a new system of undergraduate advising.

✯ Successes and Strategies

Developing the online matrix of competencies and skills being taught across the curriculum was useful in identifying gaps for the faculty.

After implementing matrix approach, we have a more cohesive curriculum that stresses repeated exposure, expected mastery of key skills necessary for conducting research, reporting research (oral and written), analyzing data and designing research plan.

Focused on streamlining student learning outcomes and methods of assessment. This has resulted mainly in attempts to integrate more critical thinking, writing, presentation skills, and teamwork into existing classes. There has also been a concerted effort to encourage faculty to engage classes with more active/experiential learning exercises.

Strategies that worked well were implementing changes incrementally, and as part of planning that had to be done for external assessment.

Retreat was very successful.

After a retreat, faculty generally agreed on the overall goals, concepts and skills that need to be emphasized throughout the curriculum. We did not need to add courses, but we agreed to make courses more consistently emphasize important concepts and skills as a student progresses.

The NAGT TWP workshop was very much a success — we had a lot of great discussions and ideas about possible course offerings — an outcome of these discussions was the development of the Violent Earth course.

We've had the most success implementing change when a small group, or even just one faculty member, is willing to take the lead (rather than top-down).

Identifying a core group faculty to define goals and implement strategies is important.

Try to overcome entrenched ideas regarding what constitutes a "real" BS degree in geology; traveling workshops may help with this.

Creating a poster-size, color-coded map of faculty course offerings and enrollments going back four years was useful for bringing a new Chair up to speed on curriculum and scheduling. It helped identify inefficiencies and how to shift resources to new curriculum initiatives.

To increase recruitment and retention one department developed two tracks that lead to a degree in Geology. The "Career Track" will prepare students for either graduate school or employment in the field of geosciences. The "General Track" provides the student with a solid background in the geosciences, however it is probably lacking in some of the cognate courses (specifically Calculus, and less rigorous Chemistry and Physics) that a graduate school or geoscience employer may be looking for. This second track was developed for students who enjoy geology but may end up working outside of the geosciences.

In an effort to overcome the limited math and computer skills of undergraduates, we are working to include examples in every course to show how these skills are important and applied in Geological Sciences. Pairing courses throughout the undergraduate trajectory is effective because students see the connections between courses and can spend more time as teams working on projects.

To aid in career readiness, students in specific geoscience courses are required to complete a 'signature assignment' to be uploaded to an *ePortfolio*. At the end of each semester, student ePortfolio 'signature assignments' are assessed on several specific criteria. A small financial incentive by the University increased faculty willingness to implement the ePortfolio requirement into their courses.

After the State's decision to provide free 2YC tuition to all new high school graduates and adults who have not completed a college degree, we made great strides in developing 2YC to 4YC geology transfer pathway/academic map with area schools.

△ Challenges and Roadblocks

Convincing certain faculty that they might need to change their courses to include important competencies/skills can be difficult.

Problems with obstruction from senior faculty reluctant to relinquish control that they have historically exercised.

Difficulty to get faculty to make concrete written plans and continually had to show by example how intertwined our curriculum matters are and need to be discussed across department not in discipline specific meetings.

Difficult to get faculty to pay attention or focus on curricular issues, particularly assessment.

We are a small program (4 faculty) and one faculty member is a reluctant change agent.

The new Chair had trouble motivating senior, tenured faculty. By enlisting the help of their undergraduate coordinator, who knows the curriculum in detail, he was able to fill in the blanks and create a matrix.

Progress has been hampered by a lack of faculty buy-in and a shrinking faculty size. Older faculty are traditionalists insisting on status quo creating obstacles to change and untenured faculty are afraid to speak up.

Personnel turnover and a necessary focus on hiring took time.

No internal accommodation or reward for any of these efforts. The Geoscience Program Coordinator position does not come with any course release and so any time spent on program initiatives has to be extracted from time spent on a high teaching load (3 courses per semester) or from the Coordinator's research program.

Progress has been hampered by the overriding crisis of the budget and higher education funding situation: ~70% budget cuts and my undergrad program has seen a 50% decline in majors."

We are just beginning to hit roadblocks (2020 update) because of budget cuts in the past year, loss of staff and other financial uncertainties.

Concerns regarding differential teaching implications on promotion and tenure of junior faculty. Make sure metrics for P/T are well understood.

Limited math and computer skills in our undergraduates, so introducing math and computation is difficult. We are working to include examples in every class to show how math and computational skills are important and applied.

We need to do a better job of communicating to students what our expectations are.

One difficulty we face is collecting data about employment possibilities that we can use to induce administration support and serve to attract potential students and their parents to our program.

The University Marketing office has been uninterested in working with programs.

Plans were slowed by continued governance problems associated with a poorly conceived/planned merger that occurred 12 years ago. Convinced the faculty to re-think our entire charter.

No time or resources to increase participation with 2YC campuses.

Advice

Utilize an undergraduate advisor to let students know what are the overall goals of the obtaining a degree in Geological Sciences. They need to see the big picture. Also, career advising and access to internships are key to opening up students' horizons.

Be patient and be persistent. Although we are making progress we are still at least a year away from having concrete evidence of progress.

I think the best piece of advice is to be patient, but insistent that changes can improve our offerings and be beneficial to our students and to our program. In times of budget problems, these kinds of changes can be program savers.

In my experience, making slow and steady changes incrementally was received best by most faculty.

For major changes in governance, start early with a rational group of faculty to gather opinions. Don't let a few individuals dominate the conversation in a general faculty meeting. Some may take a position that they think is supported, but in reality the junior faculty are afraid to confront those individuals. Set guidelines for behavior in faculty meetings (at least that has become an issue for my faculty).

Take it slow and spend the time to get faculty buy-in. Before you start, figure out how to overcome entrenched ideas regarding what constitutes a "real" BS degree in geology; traveling workshops may help with this. Be prepared to do a lot of background research and bring that to the table before you engage your faculty in discussions involving major changes. Incentivize things that you can as Chair.

Be more realistic about expectations.

Paving the way for regular discussion among faculty is critical to having a functional department. Regular discussions on teaching can dispel preconceptions about what your colleagues may be doing — they can also lead to greater cohesion and integration in the curriculum, along with new course ideas and teaching strategies. As we move toward figuring out how to operate under new challenges (whether that be a financial crisis, a global health crisis, or, both), I think this communication and teamwork will be critical to keeping our department afloat in uncertain times.

If you are at an institution where merit pay increases are possible, you can encourage change by increasing the weight of various faculty activities in your annual merit review process. If you make "uses engaged learning practices" worth 20–30% of someone's teaching evaluation score in your merit review, and if you wield that evaluation sincerely and critically, you will get people to start using engaged learning practices.

A good relationship with the administration is useful: be positive, and serve on committees. But, also be careful and be prepared for unforeseen changes. Remember to emphasize the importance of the liberal arts, if yours is a liberal arts college.

Communicate curricular decisions to the Dean and develop a budget to support implementation. Deans who are not geoscientists may not be understanding of why the faculty is proposing changes, so on-going communication with the Dean about decisions being made is important.

Courtesy of the Jackson School of Geosciences, University of Texas at Austin

The twenty-one doctoral R1 institutions range in faculty size from 5–51:

Sixteen doctoral R1 universities submitted 2017 progress reports. Four of those institutions provided 2019 or 2020 updates, and five additional institutions provided new 2019 progress reports. All but one of the institutions addressed **curriculum redesign,** through a range of approaches. Most of the institutions used the curriculum matrix to describe the current program and design a set of recommendations outlining curriculum reform. Changes include elimination of redundancies in curricula, discussion of revising sequence along lines of core-competency goals, implementation of competency-based BS and BA curricula, implementation of a common two-year core curriculum for three undergraduate degree programs and other revision of degree programs (BS Geology and BS Geophysics.). A few departments decided on the big picture reforms their curriculum needed first, designed a curriculum around those, and then did a course-by-course matrix later to show where skills and concepts were being covered. Strategies for faculty buy-in: modest tweaks, plans for faculty to review for consideration, discussion, amendment as necessary, and approval, eventual implementation. Additional approaches to curriculum design include submission of a NSF pathways grant, plans for future work on assessment plans, and creation of a "learning goals and outcomes" document with concepts, skills and competencies for undergraduate curriculum. Updates in 2019 include a desire among younger faculty to continue moving on curriculum reform.

Six institutions made progress on addressing **pedagogy.** These changes include emphasis on experiential learning, career-building exercises, scaffolding writing throughout the major, and field instruction. Additionally, it included providing professional development information to junior faculty and support of faculty to pursue their interests in course reform while ensuring that peer evaluators are aware of this encouragement and of the possibility of poor student evaluations along the way. Peer mentoring in active-learning classrooms, in which "new" faculty co-teach courses with seasoned instructors and then take over the course as lead instructor after 1–2 semesters, along with new peer teaching evaluation that includes teaching practices inventory.

Two institutions addressed **2YC–4YC transfers,** through meetings with local community colleges and submission of NSF Pathways grant with a primary goal of connections to 2YCs. Three institutions have future plans to address **underrepresented populations**, preparing grants for future submission aimed to increase underrepresented participations in programs, outreach, and pedagogical reform to promote success for underrepresented minorities and women, and development of plans associated with campus-wide initiatives to develop an inclusive excellence plan.

Four institutions made **changes to the majors**, reducing the number of majors and options, creating new programs (BS in Oceanography), merging and aligning 3 degrees into one, and eliminating majors. Four institutions worked to **change courses**, with new courses/curricula first introduced as electives prior to full enrollments during the following year, revising specific courses (separating Field Mapping from Structural Geology), renewing the non-majors core curriculum through addition of interdisciplinary courses, adding "intersession" courses to promote student progress in the degree, and developing online courses/formats for 100-level, non-major courses. Additional comments from the 2019 updates include: addressing skill gaps recognized as a result of matrix-based curriculum redesign, and how to assess whether curriculum changes are leading to better student learning outcomes.

One of the institutions does not have a geoscience program, but proposed and was approved for a new interdisciplinary minor in Environmental and Sustainability Studies that works for six colleges and is working towards making this a major. A wide range of science electives and a mandatory interdisciplinary methods course has contributed to making the minor successful.

✯ Successes and Strategies

We brought Dave Mogk in to present a seminar and afternoon workshop on undergraduate Geosciences curriculum planning and content. Initiated a special curriculum committee to make recommendations on courses, pre-requisites, learning objectives, and alignment.

The SERC pages on retreat planning and Backward Curriculum Design are very helpful for planning a retreat.

Making progress on curriculum reform required an all-day retreat, across campus with catered meals. That bought the eight hours x 40 faculty required to get the basic data, which a smaller group of staff and leadership could then use to create the final matrices for all three undergraduate tracks. After identifying gaps, courses have been redesigned to emphasize the competencies and skills identified in the workshops.

We are now using the matrices to help revision our curriculum. We have long thought we were good at hydrology, but the matrices have identified that a student could get through our curriculum without encountering much hydrology, which is reflected in the poor performance of many students during the Hydrology Section of Field Camp. We are now exploring ways to enhance fluid-related content throughout the curriculum.

We really like the matrices, and they are resulting in action.

Running all of this through a committee using a matrix "strawman" allowed us to get beyond the minutia of faculty concerns that can grind these efforts to a halt. One-on-one meetings also allowed us to get information quickly and in an environment that wasn't threatening to the faculty.

Faculty buy-in may be facilitated by progressing slowly enough to allow feedback. For example, for two weeks the department filled two walls with posters listing their learning goals. This allowed everyone an opportunity to provide feedback.

Department chair guided debate, built consensus, and got faculty approval to implement most of the proposed change.

The most successful strategy after getting faculty approval to implement the agreed upon curriculum changes was to delegate the University mandated submission process to a single faculty member and grant her a teaching release. The submission process was a lot of work and it could not have happened without the dedicated attention of one person.

Changes made from previous external review are well established and faculty are now considering how to respond to changing workforce demands by establishing specific tracks within our degree that allow students to tailor their degrees earlier in the areas of Petroleum Geosciences, Environmental Geosciences, and Geology.

One of our aims has been to increase quantitative skills for our graduates. We are trying to do this by introducing more quantitative activities across a range of courses. We are also about to launch our own Quantitative Geosciences course. The course will try to address the problem we have that a lot of geology students feel that are "bad at math". We aim to teach them the math and statistical skills they need for the job market today.

After the University Courses & Curriculum Committee blocked the proposed addition of a Data Analysis course in our department, the solution was to instead require a similar course taught by a sister department that had already been approved by the University. While the data sets will not be geological the teaching load is shifted from us to another department.

I have made it a regular practice to discuss approaches used in teaching, both at introductory and upper level, with all faculty during annual reviews and mid-year update discussions. When I hear something new and innovative from one faculty member, I share that with the other faculty members as I talk to them so they can consider whether it might work for them. It means they don't have to go out looking for new ideas (saving them time), but it also makes it clear to them that their colleagues might be doing some more innovative and imaginative things in their classes than they are. As a result, more faculty are engaging in active learning approaches.

Time (months, years) and patience. One step at a time worked for us, so it took us MONTHS to get through, but this takes time. You cannot rush this, faculty have to see the results, go away and ponder, then come back.

Having open dialog and getting all stakeholders involved in the process, but we still have a long way to go.

Enlisting a core group of faculty members who could serve as representatives from different parts of the department (bottom-up instead of top-down was key) and taking time to process best practices. The core group should be impassioned by the objectives, but has to be willing to compromise in the end if necessary.

Engage faculty. I am lucky that our faculty, at least a group of them, were relatively willing in participating these activities, once they understand that these activities will strength our department's standing and ultimately benefit the students and fulfill our long-term vision. We did have a retreat and one of the two major agenda items of the retreat was on teaching.

Empower faculty to champion. We now have a very unlikely 'hero' leading the efforts on developing the concept-skill metric and revamping our freshman-level classes. I said "unlikely' because that faculty had always put research above everything else.

We noticed that we had placed much of the transferrable skills in specific classes (e.g., writing, in one course titled Scientific Communication). This meant that students would get these skills once, but maybe not again. We are now alpha-testing infusing these transferrable skills into the majors courses (e.g., writing exercises, taught by a second instructor who specializes in writing, etc.)

Just getting the faculty to separate conceptually the curriculum issues from "I want to tell you how to teach" was a major success. The discussions were good.

We got faculty into the classroom to break down the mystique of "flipped" classrooms. It's not complicated or scary but the buzzwords put people off. Showing them and making them part of it breaks down these preconceived notions quickly!

Peer mentoring in active-learning classrooms has been successful. "New" faculty co-teach these courses with seasoned instructors and then take over the course as lead instructor after 1–2 semesters. Introduction of a new graduate training course with specific training in active learning greatly improved TA support in these types of courses.

Early career faculty are most receptive, in a one-on-one setting, to fairly simple, low-cost approaches that might enhance their classes.

We are slowing drawing more faculty into transformed introductory courses and these experiences appear to "trickle" up to their major courses. The mentoring approach used within the transformed courses were very successful. Student evaluations for these instructors were immediately high without any of the dips seen with other implementation strategies. The TA training course was successful and is going to be expanded this year to include more strategies for student success (time management, etc.).

New set of interdisciplinary courses for 1st and 2nd years using innovative instructional methods such as the flipped classroom and blended learning. Other faculty have seen the effectiveness and have started implementing these in other courses. As part our assessment activities, faculty members discuss the effectiveness of newer teaching methods compared with that of the older teaching methods.

Used university resources. Our Center for Teaching Excellence has an implementation grant in STEM that allows for some support for faculty who transform courses.

Successful participation in SAGE workshop (Supporting and Advancing Geoscience Education) where selected pedagogical methods were summarized and demonstrated.

Changes in leadership and new hires was a catalyst for change for departments.

Working with the deans and trying to get more resources. We successfully gained a new tenure-track FTE on geoscience education through a special opportunity hire. We may also get additional TA support for the new freshman-level courses we are developing. Deans and upper administration were happy with our willingness and efforts; how can they not support?

Social media is an effective tool for reaching students. We have an up-to-date Geoscience Undergrad Facebook page designed to help facilitate the academic and career goals students of majoring in Geology, Geography, and Environmental Science and Policy.

Collaborated with Science Ed to submit a NSF Pathways grant (to connect with local 2YCs).

Working with science colleagues (Biology and Chemistry) mostly worked in establishing the interdisciplinary Minor in Environmental and Sustainability Studies, because we had a mutual interest in getting students enrolled in these classes.

△ Challenges and Roadblocks

A fundamental split among our faculty on what the future of geosciences should be. Some cannot see that geosciences need to be any different than what geology has been for decades. They want nothing added, nothing removed, nothing changed. They will concede adding something non-traditional (such a courses on climate or water!) as electives, but never at the expense of something they regard as fundamental. Other faculty feel that geosciences is changing and needs to change to meet current and future needs, and are trying to push for a curriculum that better connects geosciences with sustainability.

Senior faculty have been so recalcitrant that some have even opted to phase in their retirement so as not to be bothered with discussions of curriculum and updated pedagogy.

Some faculty are not interested in critically evaluating teaching and curriculum. While they may want to do a good job individually in their courses, there is little interest in thinking holistically about the undergraduate program and how its pieces fit together. There is no mechanism to enforce adoption engagement by faculty. Instead the effort is driven by a small group (~6 out of 23).

There was the expected overenthusiasm for expanding every part of the curriculum as too important to miss. We did projections of how much space was in the degree, how many courses we could reasonably teach, how much lab space we had, in order to inject reality.

The Mogk matrix approach bogged us down in an overly detailed process at the expense of the big-picture reforms our curriculum needed.

Difficult to keep large faculty committee focused and productive.

There was some initial chaos when these ideas were initially vetted in a faculty meeting to the entire faculty.

I developed the competencies matrix to identify areas of strength and weakness within our program. The faculty felt, we were covering the competencies we wanted. However, beyond that, I have not been able to get faculty to engage with this approach and the matrix has largely been ignored since

---

Convening a committee to establish program-wide learning goals and outcomes went fairly well. But I've realized that we have no mechanism in the department to enforce anything else. There is no way to force putting these learning goals in to any course, and no mechanisms to enforce assessment of these. The biggest problem to achieving our action plan is faculty inaction. This endeavor is largely not a priority for the majority of our faculty.

No matter how good the idea, it always comes down to politics and relationships. Everyone agreed it was a great idea, but everyone said no, not until we spoke with and got the approval of x, y, or z. Meeting-after-meeting, presentation-after-presentation, briefing-after-briefing, we finally got decision-makers to agree that it was not only a great idea, but that we also should go ahead and pursue it.

Have a vision and share that regularly with the faculty, but don't expect miracles.

Early-career faculty feel tremendous pressure to do their research, and so investing too heavily in pedagogical experiments ranks low. However, these are the faculty who are most receptive when I share with them (one-on-one) a fairly simple, low-cost approach that might enhance their classes.

The strongest impediment is senior faculty thinking that this process will tell them how to teach their material. "How to Teach" is a different issue than "What to Teach."

Long-term individual faculty ownership of particular course that may be modified by this process.

No interest from faculty in addressing when teaching methods successes and failures discussed at faculty meetings.

I also tried to establish working groups around several of our introductory-level courses as a way of sharing best practices, but those faculty who would probably have benefited most by participating were least interested and complained that they did not want to be told what they could and could not teach. Getting well-established faculty to change their approach can be extremely difficult. They want to keep doing what they've always done.

A challenge that we think the geoscience community will be facing: a mismatch between our hopes and dreams vs. the reality of how difficult it is for students to transition from just absorbing knowledge to actively engaging in the process of generating new knowledge.

Getting faculty to teach in areas that may be higher enrollment.

University Committee questions our proposed course in Data Analysis. Solution was to require a course in another department.

One of the biggest roadblocks we have as a department is inadequate facilities and resources to fix them.

Time (2 departments).

Merger of programs: Unit head and faculty recalcitrance, not because the idea isn't good, but because of a perception of loss of control and loss of resources like TA-ships, which would be shared by merged program. Our program is a traditional one with easily identifiable core-competencies and the degree we wanted to merge with is elective heavy and the faculty in charge were not willing to list and rank the core competencies they address now, and how those might be modified in the future. Roadblock is working out the administration of this first and second year common curriculum and the buy-in from the three departments (because of the perceived loss of program control over curriculum).

Advice

I've found the SERC pages on retreat planning and Backward Curriculum Design very helpful as I plan our upcoming retreat.

The all-day retreat was REALLY worthwhile. We spent some time explaining the matrix approach, and getting us all on the same page with regards to the Likert scaling. Then we broke into groups, which were fluid so people could join-and-leave multiple groups and real-time feedback could help us all get/stay on the same page.

Aim high but expect setbacks. Be patient. One step at a time.

Show faculty the results of the survey from the Summits and Geoscience Employers workshop.

Let everyone (faculty, staff, admin) feel like they are part of the entire process from the first steps of the initial planning efforts. It takes longer in the beginning, but having everyone on board from the start is worth the (sizable) initial investment to (hopefully) avoid issues down the road.

Don't discuss, plan, or dictate anything without frequently asking for every faculty member's feedback.

Get a core group of faculty that can serve as representatives to various parts of the department; take your time to think it all through and do it right; consider logistics right from the outset. The core group needs passion for the project, but has to be willing to compromise in the end if necessary for agreement.

Try to avoid having the process drawn out; it runs the danger of losing momentum. Incorporate reports to the general faculty on in-progress committee deliberations as a part of the general approach, to keep them apprised, engaged, and able to give committee members feedback (though this would require dedicating a lot of faculty meeting time to the topic).

Patience, patience, patience, one step at a time. All faculty think this was very good for the Department. They might not have "enjoyed" the process, but they realize we made good changes and used the process to motivate the changes. We did NOT look for "overlap". I think discussion of "overlap" would be a "hot button" issue. In addition, some faculty are now passing syllabi a back and forth between prerequisites and courses.

Make sure that there is some mechanisms in place for driving and enforcing your proposed changes.

Slow process. Have committee of earlier-career faculty from diverse fields to build plan. We did start with learning outcomes, knowledge and skills, and these did help the process. Altogether took four+ years. Could have been faster if no other distractions, but three years would have been the fastest from start to finish.

Recognize that there are alternative approaches to a Mogk matrix analysis of current classes. Get the faculty to decide on the end result of reform and work from there.

Connecting personally and individually with professors built support and created momentum to continue the process. Get faculty into the classroom to break down the mystique of "flipped" classrooms.

As student research experiences are added to the curriculum, the intent is that students are learning what it means to be a scientist. Be cautious of a mismatch between the intent of the research experience and the reality of how difficult it is for students to transition from absorbing knowledge to actively engaging in the process of generating new knowledge.

Check out university/college resources as many have teaching centers or other resources that help faculty learn new teaching methods.

Don't downplay or discount administration or curricular demands at faculty meetings, this is the boat we are in, we must sail it together.

RUBRIC FOR QUANTIFYING PROGRESS MADE TOWARD ACTION PLAN OBJECTIVES

The following was applied to quantify the progress a department has made toward accomplishing their individual Action Plan objectives:

- 10–20 — participant shared their plan with colleagues but failed to get faculty onboard
- 20–30 — participant/department did one or more of the following: initiated analysis of existing curriculum; began to produce matrices; provided resources for improving pedagogy; identified specific steps for implementation of plan objectives
- 30–40 — completed department-specific matrices or other plan elements but no tangible outcomes; progress stalled due to external factors (no support of leadership, budget constraints, hiring freeze, etc.)
- 40–50 — plan for implementation defined and initiated (possibly awaiting University approval)
- 50–60 — tangible progress in either curriculum redesign (successfully implemented in a course or two), pedagogy (faculty slowly adopting new ideas, attending workshops, transforming courses) or other major plan elements
- 60–100 — reflects the scale to which new curriculum, courses, pedagogy and other objectives are being adopted in the department

If the primary objective was curriculum and/or pedagogy, but the department made significant progress addressing an increase in underrepresented groups, K–12 teaching, 2YC–4YC, change in majors or additional courses, they increased by 10–20.

Appendix D: Participants

Over 1,000 members of the geoscience community contributed to the development of the vision for the future of geoscience undergraduate education in the United States. The level of input to the development of the vision was measured, on the academic side, by institution type (Table D-1), and on the employer side, by industrial sector (Table D-2). The level of input reflects the total number of contribution points — individuals attending summits, responses to surveys, or progress reports submitted.

Additionally, we have listed all academic institutions which had representatives contribute to the vision, as well as employers who were engaged in the process too. The list does not include participants from additional academic institutions and employers who provided anonymous responses to the survey.

Institution Type	Inputs to the Vision
Community Colleges	11%
Bachelor's-granting	33%
Terminal Master's-granting	6%
Non-US Doctoral-granting	1%
Research 1 Doctoral-granting	31%
Other Doctoral-granting	18%
All Doctoral-granting	50%

Table D-1: Magnitude of inputs (summit attendance, survey responses, and progress reports) by institution type.

Employer Sector	Input to the Vision
Other energy	1%
Other consulting	1%
Education	2%
Environment/hydro/eng.	9%
Mining/minerals	14%
Petroleum	26%
Petroleum consulting	3%
Reinsurance	1%
Weather/climate	3%
Federal agencies	19%
Museums	2%
Non-profits	7%
State agencies/surveys	3%
Professional societies	14%

Table D-2: Magnitude of inputs (summit attendance, survey responses, progress reports) by employment sector

All courtesy of the Jackson School of Geosciences, University of Texas at Austin

ACADEMIC INSTITUTIONS

Adrian College
Albion College
Amherst College
Angelo State University
Arizona State University
Arkansas Tech University
Augustana College
Austin Community College
Ball State University
Bare Mountain Community College
Baylor University
Bemidji State University
Benedictine University
Blinn College — Bryan Campus
Bloomsburg University
Boise State University
Boston College
Bowdoin College
Bowling Green State University
Brigham Young University — Idaho
Bucknell University
Buena Vista University
California State University Chico
California State University, Bakersfield
California State University, East Bay
California State University, Fullerton
California State University, Long Beach
California University of Pennsylvania
Cape Fear Community College
Carleton College
Centenary College of Louisiana
Central Michigan University
Central Oregon Community College
Central Washington University
Central Wyoming College
Centralia College
City College of San Francisco
City University of New York
Clemson University
Colby College
College of William & Mary
Colorado School of Mines
Colorado State University
Columbus State University
Community College of Rhode Island
Cornell University
Cuyahoga Community College
Dartmouth College
Daytona College
Del Mar College
Drexel University
Dutchess Community College
East Carolina University
East Tennessee State University
Eastern Carolina University
Eastern Connecticut State University
Eastern Illinois University
Eastern Washington University
Edinboro University of Pennsylvania
El Paso Community College
Elgin Community College
Essex County College
Everett Community College
Fitchburg State University
Flagler College
Florida Agricultural and Mechanical University
Florida Atlantic University
Florida Gulf Coast University
Florida International University
Florida State College at Jacksonville
Florida State University
Fort Hays State University
Fort Valley State University
Front Range Community College
George Mason University
Georgia Southern University
Grand Valley State University
Green River College
Gustavus Adolphus College
Hamilton College
Hardin-Simmons University
Highline Community College
Hill College
Hope College
Humboldt State University
Idaho State University
Illinois Valley Community College
Indiana University of Pennsylvania
Indiana University/Purdue University
Indiana University Bloomington
Iowa State University
Ivy Tech Community College of Indiana
James Madison University
Kansas State University
Kent State
Kent State University
Lake Superior State University
Lamar University
Lamont-Doherty Earth Observatory of Columbia University
Lane Community College
Laurentian University
Lawrence University
Long Island University Post
Lord Fairfax Community College
Louisiana State University
Lyndon State College
Macalester College
Marshall University
Massachusetts Institute of Technology
Mesa Community College
Mesalands Community College
Metropolitan Community College — Kansas City
Michigan State University
Michigan Technological University
Mid-Continent University
Middle Tennessee State University
Millsaps College
Mississippi State University

Montana State University
Montana Technical University
Mt. San Antonio College
Murray State University
Muskegon Community College
National Taiwan Normal University
Nevada State College at Henderson
New Jersey City University
New Mexico Institute of Mining and Technology
New Mexico State University
North Carolina State University
Northeastern University
Northern Arizona University
Northern Illinois University
Northern Virginia Community College
Northland College
NorthWest Arkansas Community College
Northwest Florida State College
Northwest Missouri State University
Northwestern University
Oberlin College
Occidental College
Ohio State University
Old Dominion University
Olivet Nazarene University
Oregon State University
Pacific Lutheran University
Pasadena City College
Penn State Brandywine
Penn State University (Geoscience)
Penn State University (Meteorology)
Plymouth State University
Pomona College
Portland Community College
Prairie State College
Princeton University
Purdue University
Radford University
Rice University
Rocky Mountain College
Saint Louis University
Salem State University
San Francisco State University
San Jose City College
San Jose State University
Santa Rosa Junior College
Scottsdale Community College
Scripps Institution of Oceanography, University of California, San Diego
Shawnee State University
Skagit Valley College
South Dakota School of Mines
South Mountain Community College
Southwestern Illinois College
St. Petersburg College
Stanford University
Stockton University
Stony Brook University
Sul Ross State University
SUNY Buffalo State College
SUNY Oneonta
SUNY Oswego
SUNY Potsdam
Syracuse University
Tacoma Community College
Tarrant County College — Northeast Campus
Temple University
Tennessee State University
Tennessee Tech University
Texas A&M University (Atmospheric Sciences)
Texas A&M University (Geology & Geophysics)
Texas A&M University (Oceanography)
Texas Christian University
Texas Tech University
Thomas Nelson Community College
Towson University
Trinity University
Tufts University
University at Albany
Universidad de Chile
University of Adelaide
University of Alabama
University of Alaska Fairbanks
University of Alberta
University of Arizona
University of Arkansas at Little Rock
University of Arkansas at Pine Bluff
University of British Columbia
University of Calgary
University of California, Berkeley
University of California, Davis
University of California, Los Angeles (UCLA)
University of California, San Diego
University of California, Santa Cruz
University of Canterbury, New Zealand
University of Colorado Boulder
University of Connecticut
University of Delaware
University of Denver
University of Florida
University of Georgia
University of Hawaii at Manoa
University of Houston
University of Illinois at Chicago
University of Illinois at Urbana-Champaign
University of Iowa
University of Kansas
University of Kentucky
University of Louisiana at Lafayette
University of Manitoba
University of Massachusetts Amherst
University of Massachusetts Boston
University of Miami
University of Michigan — Dearborn
University of Minnesota
University of Minnesota Morris

University of Missouri
University of Montana Western
University of Mount Union
University of Nebraska Omaha
University of Nevada, Reno
University of New Mexico
University of North Carolina at Pembroke
University of North Carolina at Charlotte
University of North Dakota
University of Northern Colorado
University of Northern Iowa
University of Ohio
University of Oklahoma
University of Pennsylvania
University of Puerto Rico — Mayaguez Campus
University of Rhode Island
University of Rochester
University of Saskatchewan
University of South Alabama
University of South Carolina
University of South Florida
University of Southampton
University of Southern California
University of Southern Indiana
University of Southern Maine
University of Tennessee
University of Tennessee at Chattanooga
University of Tennessee at Martin
University of Texas at Austin
University of Texas at Dallas
University of Texas at El Paso
University of the Pacific
University of the South
University of Toledo
University of Utah
University of Washington Bothell
University of Waterloo
University of West Florida
University of West Georgia
University of Wisconsin — Madison
University of Wisconsin — Parkside
University of Wisconsin — Richland
University of Wyoming
University Western Michigan
Utah State University
Vanderbilt University
Washington State University
Waubonsee Community College
Wayne State University
Weber State University
Wenatchee Valley College
Western Carolina University
Western Colorado University
Western Illinois University
Western Kentucky University
Western Michigan University
Western New Mexico University
Western Washington University
Wharton County Junior College
Whatcom Community College
Wheaton College
Wittenberg University
Yale University
York College (City University of New York)

EMPLOYERS

AAAS (Science & Technology Policy)
AIR-Worldwide
American Geophysical Union
American Geosciences Institute
American Meteorological Society
Anadarko Petroleum
ARCADIS U.S., Inc.
Aztec Geoscience Inc.
Bayhorse Silver Inc.
BHP Billiton
Blue Moon Exploration Company
Chevron Energy Technology Company
Cleveland Museum of Natural History
Conoco Phillips Company
CONSOL Energy Inc.
Consortium for Ocean Leadership
DeGolyer & MacNaughton
Draper Aden Associates
Encino Energy
Endeavour Silver
Environmental Protection Agency (EPA)
EOG Resources
Eriksson Associates
ExxonMobil Exploration Company
ExxonMobil Upstream Research
Freeport McMoRan
Geochimica, Inc.
Geological Society of America
Geological Survey of Alabama
Geological Survey of New South Wales
Hess Corporation
Hunter Dickinson Inc.
IBM and the Weather Company
IMDEX/Cascabel
Institute of Marine Engineering, Science and Technology
Integral Consulting
IOS Services Géoscientifiques
IRIS Consortium
Jackson Exploration LLC
Jacobs Environment/Hydro/Engineering
Jupiter Intelligence
Langan Engineering and Environmental Services
Leigh Freeman Consultancy
LLOG Exploration Company
Louisiana Department of Environmental Quality
Louisiana Geological Survey

Marathon

Martineau Petroleum

Murchison Minerals, Inc

NASA Johnson Space Center, Astromaterials Research and Exploration Science (ARES)

NASA Goddard Flight Center (Climate)

NASA Goddard Space Flight Center (Meteorology)

NASA Headquarters (Earth Sciences)

NASA Jet Propulsion Laboratory (Geophysics and Planetary Geosciences)

NASA Jet Propulsion Laboratory Science Division

National Academies of Sciences, Engineering, and Medicine, Ocean Studies Board

National Earth Science Teachers Association

National Oceanic & Atmospheric Administration, Education

National Oceanic & Atmospheric Administration, Northeast Fisheries Science Center

National Oceanic & Atmospheric Administration, Research Fisheries

National Science Foundation, Antarctic Science Section

National Science Foundation, Atmospheric and Geospace Sciences

National Science Foundation, Earth Sciences

National Science Foundation, Education & Diversity (GEO/OAD)

National Science Foundation, Education and Cross Disciplinary Activities Program (GEO/AGS)

National Science Foundation, Geoscience Directorate

National Science Foundation, Geospace Science

Courtesy of the Jackson School of Geosciences, University of Texas at Austin

National Science Foundation, Graduate Education

National Weather Service, National Center for Environmental Prediction

National Park Service

New World Associates

Newmont Mining Corporation

Orogen Royalties Inc.

Paleontological Research Institution

Paleontological Research Institution, Museum of the Earth, Cayuga Nature Center

Pebble Creek Mining Ltd

Peregrine Petroleum LLC

Powell Xploration LLC

Rock Whisperer LLC

Royal Dutch Shell

Royal Exploration Co Inc.

S&ME, Inc.

Scripps Oceanographic Institute

SEPM — Society for Sedimentary Geology

Society of Exploration Geophysicists

Soil Science Society of America

Spire Global

State of Nevada Department of Transportation

Swiss Re

Talisman

Terracon Consultants

Tetra Tech

Total E&P Research & Technology USA, LLC

U.S. Department of Energy

U.S. Geological Survey, Education

U.S. Geological Survey, National Minerals Information Center

U.S. Nuclear Regulatory Commission

UNAVCO

University of California Museum of Paleontology

URS Corp

USDA-OCE — Climate Change Program Office

Violet Energy

Virginia Institute of Marine Science

WeatherCall Services

Weathernews Inc.

Woods Hole Oceanographic Institution

Courtesy of the Jackson School of Geosciences, University of Texas at Austin

Made in the USA
Las Vegas, NV
03 November 2026

57421081R00109